R 4 Data Science Quick Reference

A Pocket Guide to APIs, Libraries, and Packages

Second Edition

Thomas Mailund

Apress®

R 4 Data Science Quick Reference: A Pocket Guide to APIs, Libraries, and Packages

Thomas Mailund
Aarhus, Denmark

ISBN-13 (pbk): 978-1-4842-8779-8   ISBN-13 (electronic): 978-1-4842-8780-4
https://doi.org/10.1007/978-1-4842-8780-4

Managing Director, Apress Media LLC: Welmoed Spahr
Acquisitions Editor: Steve Anglin
Development Editor: James Markham
Coordinating Editor: Mark Powers

Cover designed by eStudioCalamar

Cover image by natanaelginting on Freepik (www.freepik.com)

Distributed to the book trade worldwide by Apress Media, LLC, 1 New York Plaza, New York, NY 10004, U.S.A. Phone 1-800-SPRINGER, fax (201) 348-4505, e-mail orders-ny@springer-sbm.com, or visit www.springeronline.com. Apress Media, LLC is a California LLC and the sole member (owner) is Springer Science + Business Media Finance Inc (SSBM Finance Inc). SSBM Finance Inc is a **Delaware** corporation.

For information on translations, please e-mail booktranslations@springernature.com; for reprint, paperback, or audio rights, please e-mail bookpermissions@springernature.com.

Apress titles may be purchased in bulk for academic, corporate, or promotional use. eBook versions and licenses are also available for most titles. For more information, reference our Print and eBook Bulk Sales web page at http://www.apress.com/bulk-sales.

Any source code or other supplementary material referenced by the author in this book is available to readers on GitHub (https://github.com/Apress). For more detailed information, please visit http://www.apress.com/source-code.

Printed on acid-free paper

Table of Contents

TABLE OF CONTENTS

About the Author

Thomas Mailund is an associate professor in bioinformatics at Aarhus University, Denmark. His background is in math and computer science, but for the last decade, his main focus has been on genetics and evolutionary studies, particularly comparative genomics, speciation, and gene flow between emerging species.

About the Technical Reviewer

Matt Wiley leads institutional effectiveness, research, and assessment at Victoria College, facilitating strategic and unit planning, data-informed decision making, and state/regional/federal accountability. As a tenured, associate professor of mathematics, he won awards in both mathematics education (California) and student engagement (Texas). Matt earned degrees in computer science, business, and pure mathematics from the University of California and Texas A&M systems.

Outside academia, he coauthors books about the popular R programming language and was managing partner of a statistical consultancy for almost a decade. He has programming experience with R, SQL, C++, Ruby, Fortran, and JavaScript.

A programmer, a published author, a mathematician, and a transformational leader, Matt has always melded his passion for writing with his joy of logical problem solving and data science. From the boardroom to the classroom, he enjoys finding dynamic ways to partner with interdisciplinary and diverse teams to make complex ideas and projects understandable and solvable.

CHAPTER 1

Introduction

R is a functional programming language with a focus on statistical analysis. It has built-in support for model specifications that can be manipulated as first-class objects, and an extensive collection of functions for dealing with probability distributions and model fitting, both built-in and through extension packages.

The language has a long history. It was created in 1992 and is based on an even older language, S, from 1976. Many quirks and inconsistencies have entered the language over the years. There are, for example, at least three partly incompatible ways of implementing object orientation, and one of these is based on a naming convention that clashes with some built-in functions. It can be challenging to navigate through the many quirks of R, but this is alleviated by a suite of extensions, collectively known as the "Tidyverse."

While there are many data science applications that involve more complex data structures, such as graphs and trees, most bread-and-butter analyses involve rectangular data. That is, the analysis is of data that can be structured as a table of rows and columns where, usually, the rows correspond to observations and the columns correspond to explanatory variables and observations. The usual data sets are also of a size that can be loaded into memory and analyzed on a single desktop or laptop. I will assume that both are the case here. If this is not the case, then you need different, big data techniques that go beyond the scope of this book.

The Tidyverse is a collection of extensions to R: packages that are primarily aimed at analyzing tabular data that fits into your computer's memory. Some of the packages go beyond this, but since data science is predominately manipulation of tabular data, this is the focus of this book.

The Tidyverse packages provide consistent naming conventions, consistent programming interfaces, and more importantly a consistent notation that captures how data analysis consists of different steps of data manipulation and model fitting.

1

© Thomas Mailund 2022
T. Mailund, *R 4 Data Science Quick Reference*, https://doi.org/10.1007/978-1-4842-8780-4_1

The packages do not merely provide collections of functions for manipulating data but rather small domain-specific languages for different aspects of your analysis. Almost all of these small languages are based on the same overall "pipeline" syntax introduced with the `magrittr` package. The package introduces a syntax for sending the output of one function call into another function call. It provides various operators for this, but the most frequently used is `%>%` that gives you an alternative syntax for writing function calls:

```
x%>%f()%>%g()
```

is equivalent to

```
g(f(x))
```

This syntax was deemed so useful that a similar operator was introduced into the main R language in version 4.1, `|>`, so you will now also see syntax such as

```
x |> f() |> g()
```

I will, when it is convenient, use the built-in pipe operator `|>` in code examples, but I will leave `magrittr` syntax until after Chapter 6 where I describe the `magrittr` package.

The two operators are not identical, and I will highlight a few differences in Chapter 6. They differ in a few ways, but you can use either with the Tidyverse packages, and they are usually designed with the assumption that you use the packages that way.

A noticeable exception is the plotting library `ggplot2`. It is slightly older than the other extensions in the Tidyverse and because of this has a different syntax. The main difference is the operator used for combining different operations. The data pipeline notation uses the `%>%` or `|>` operator, while `ggplot2` combines plotting instructions using `+`. If you are like me, then you will often try to combine `ggplot2` instructions using `%>%`— just out of habit—but once you get an error from R, you will recognize your mistake and can quickly fix it.

This book is a syntax reference for modern data science in R, which means that it is a guide for using Tidyverse packages and it is a guide for programmers who want to use R's Tidyverse packages instead of basic R programming. I will assume that you are familiar with base R and the functions there, or at least that you can read the documentation for these, for example, when you want to know how the function `read.table()` behaves, you type `?read.table` into your R console. I will mention base R functions in the text when they differ from similar functionality in the Tidyverse functions. This will warn you if you are familiar with the base R functions and might have code that uses them that you want to port to Tidyverse code. If you are not familiar with base R and want a reference for the core R language, you might wish to read the book *R Quick Syntax Reference* by Margot Tollefson instead.

This guide does not explain each Tidyverse package exhaustively. The development of Tidyverse packages progresses rapidly, and the book would not contain a complete guide shortly after it is printed anyway. The structure of the extensions and the domain-specific languages they provide are stable, however, and from examples with a subset of the functionality in them, you should not have any difficulties with reading the package documentation for each of them and find the features you need that are not covered in the book.

To get started with the Tidyverse, install and load it:

```
install.packages("tidyverse")
library(tidyverse)
```

The Tidyverse consists of many packages that you can install and load independently, but loading all through the `tidyverse` package is the easiest, so unless you have good reasons to, for example, that you need a package that isn't automatically loaded, just load `tidyverse` when you start an analysis. In this book, I describe three packages that are not loaded from `tidyverse` but are generally considered part of the Tidyverse.[1]

---

[1] The Tidyverse I refer to here is the ecosystem of Tidyverse packages but not the *package* `tidyverse`, which only loads the key packages.

## CHAPTER 2

Importing Data: `readr`

Before we can analyze data, we need to load it into R. The main Tidyverse package for this is called `readr`, and it is loaded when you load the `tidyverse` package:

```
library(tidyverse)
```

but you can also load it explicitly using

```
library(readr)
```

Tabular data is usually stored in text files or compressed text files with rows and columns matching the table's structure. Each line in the file is a row in the table, and columns are separated by a known delimiter character. The `readr` package is made for such data representation and contains functions for reading and writing variations of files formatted in this way. It also provides functionality for determining the types of data in each column, either by inferring types or through user specifications.

Functions for Reading Data

The `readr` package provides the following functions for reading tabular data:

Function	File format
`read_csv()`	Comma-separated values
`read_csv2()`	Semicolon-separated values
`read_tsv()`	Tab-separated values

© Thomas Mailund 2022
T. Mailund, *R 4 Data Science Quick Reference*, https://doi.org/10.1007/978-1-4842-8780-4_2

Function	File format
read_delim()	General column delimiters[1]
read_table()	Space-separated values (fixed-length columns)
read_table2()	Space-separated values (variable-length columns)

The interface to these functions differs little. In the following text, I describe read_csv, but I highlight when the other functions differ. The read_csv function reads data from a file with comma-separated values. Such a file could look like this:

```
A,B,C,D
1,a,a,1.2
2,b,b,2.1
3,c,c,13.0
```

Unlike the base R read.csv function, read_csv will also handle files with spaces between the columns, so it will interpret the following data the same as the preceding file:

```
A,  B,  C,    D
1,  a,  a,    1.2
2,  b,  b,    2.1
3,  c,  c,    13.0
```

If you use R's read.csv function instead, the spaces before columns B and C will be included as part of the data and the text columns will be interpreted as factors.

The first line in the file will be interpreted as a header, naming the columns, and the remaining three lines as data rows.

Assuming the file is named data/data.csv, you read its data like this:

```
my_data<-read_csv(file ="data/data.csv")
```

---

[1] The read_delim() can handle any file format that has a special character that delimits columns. The read_csv(), read_csv2(), and read_tsv() functions are specializations of it. The first of these uses commas for the delimiter, the second semicolons, and the third tabs.

```
## Rows: 3 Columns: 4
##---- Column specification --------------------------
## Delimiter: ","
## chr (2): B, C
## dbl (2): A, D
##
## i Use 'spec()' to retrieve the full column specification for this data.
## i Specify the column types or set 'show_col_types = FALSE' to quiet this
message.
```

The message you get from read_csv() tells you that you can get information about the type it has inferred for each column if you use the spec() function:

```
spec(my_data)
```

```
## cols(
##    A = col_double(),
##    B = col_character(),
##    C = col_character(),
##    D = col_double()
## )
```

When reading the file, read_csv will infer that columns A and D are numbers and columns B and C are strings.

If you are happy with that, and don't want to be told about it in the future, you can use the option

```
show_col_types = FALSE:
```

```
my_data<-read_csv(file ="data/data.csv",
                 show_col_types =FALSE)
```

If the file contains tab-separated values

```
A   B   C   D
1   a   a   1.2
2   b   b   2.1
3   c   c   13.0
```

you should use read_tsv() instead.

```
my_data<-read_tsv(file ="data/data.tsv",
                  show_col_types =FALSE)
```

The file you read with read_csv can be compressed. If the suffix of the file name is .gz, .bz2, .xz, or .zip, it will be uncompressed before read_csv loads the data.

```
my_data<-read_csv(file ="data/data.csv.gz",
                  show_col_types =FALSE)
```

If the file name is a URL (i.e., has prefix http://, https://, ftp://, or ftps://, the file will automatically be downloaded.

You can also provide a string as the file object:

```
read_csv(
    "A, B, C,    D
     1, a, a,   1.2
     2, b, b,   2.1
     3, c, c, 13.0
", show_col_types = FALSE)

## # A tibble: 3 × 4
##       A B      C          D
##   <dbl> <chr> <chr> <dbl>
## 1     1 a      a        1.2
## 2     2 b      b        2.1
## 3     3 c      c        13
```

This is rarely useful in a data analysis project, but you can use it to create examples or for debugging.

File Headers

The first line in a comma-separated file is not always the column names; that information might be available from elsewhere outside the file. If you do not want to interpret the first line as column names, you can use the option col_names = FALSE.

```
read_csv(
    file ="data/data.csv",
    col_names =FALSE,
    show_col_types =FALSE
)
```

```
## # A tibble: 4 × 4
##    X1    X2    X3    X4
##    <chr> <chr> <chr> <chr>
## 1 A     B     C     D
## 2 1     a     a     1.2
## 3 2     b     b     2.1
## 4 3     c     c     13.0
```

Since the data/data.csv file *has* a header, that is interpreted as part of the data, and because the header consists of strings, read_csv infers that all the column types are strings. If we did not have the header, for example, if we had the file data/data-no-header.csv:

```
1, a, a, 1.2
2, b, b, 2.1
3, c, c, 13.0
```

then we would get the same data frame as before, except that the names would be autogenerated:

```
read_csv(
    file ="data/data-no-header.csv",
    col_names =FALSE,
    show_col_types =FALSE
)
```

```
## # A tibble: 3 × 4
##       X1 X2    X3       X4
##    <dbl> <chr> <chr> <dbl>
## 1     1 a     a       1.2
## 2     2 b     b       2.1
## 3     3 c     c      13
```

The autogenerated column names all start with X and are followed by the number the columns have from left to right in the file.

If you have data in a file without a header, but you do not want the autogenerated names, you can provide column names to the col_names option:

```
read_csv(
    file ="data/data-no-header.csv",
    col_names =c("X","Y","Z","W"),
    show_col_types =FALSE
)
```

```
## # A tibble: 3 × 4
##       X Y     Z         W
##   <dbl> <chr> <chr> <dbl>
## 1     1 a     a       1.2
## 2     2 b     b       2.1
## 3     3 c     c        13
```

If there is a header line, but you want to rename the columns, you cannot just provide the names to read_csv using col_names. The first row will still be interpreted as data. This gives you data you do not want in the first row, and it also affects the inferred types of the columns.

You can, however, skip lines before read_csv parse rows as data. Since we have a header line in data/data.csv, we can skip one line and set the column names.

```
read_csv(
    file ="data/data.csv",
    col_names =c("X","Y","Z","W"),
    skip =1,
    show_col_types =FALSE
)
```

```
## # A tibble: 3 × 4
##       X Y     Z         W
##   <dbl> <chr> <chr> <dbl>
## 1     1 a     a       1.2
## 2     2 b     b       2.1
## 3     3 c     c        13
```

You can also put a limit on how many data rows you want to load using the
n_max option.

```
read_csv(
    file ="data/data.csv",
    col_names =c("X","Y","Z","W"),
    skip =1,
    n_max =2,
    show_col_types =FALSE
)
```

If your input file has comment lines, identifiable by a character where the rest
of the line should be considered a comment, you can skip them if you provide the
comment option:

```
read_csv(
    "A, B, C,      D    # this is a comment
    1, a, a,    1.2      # another comment
    2, b, b,    2.1
    3, c, c, 13.0",
    comment ="#",
    show_col_types =FALSE)
```

```
## # A tibble: 3 × 4
##        A B     C         D
##    <dbl> <chr> <chr> <dbl>
## 1     1 a     a       1.2
## 2     2 b     b       2.1
## 3     3 c     c      13
```

You can leave a whole line as a comment, but then you want the comment character
to start to the left of that line:

```
read_csv(
    "A, B, C,   D    # this is a comment
# whole line comment
    1, a, a,   1.2    # another comment
    2, b, b,   2.1
```

```
    3, c, c, 13.0",
    comment ="#",
    show_col_types =FALSE)
```

```
## # A tibble: 3 × 4
##       A B    C         D
##    <dbl> <chr> <chr> <dbl>
## 1     1 a     a       1.2
## 2     2 b     b       2.1
## 3     3 c     c        13
```

If you have space before the comment, the function can't tell if there is an error—it looks like a line with missing columns rather than a blank line—so you will get a warning and a row with NA where the comment line was.

```
read_csv(
    "A, B, C,    D    # this is a comment
    # the indentation is a potential problem; missing columns?
    1, a, a,    1.2    # another comment
    2, b, b,    2.1
    3, c, c, 13.0",
    comment ="#",
    show_col_types =FALSE
)
```

```
## Warning: One or more parsing issues, see
## 'problems()' for details
```

```
## # A tibble: 4 × 4
##       A B    C         D
##    <dbl> <chr> <chr> <dbl>
## 1    NA <NA>  <NA>     NA
## 2     1 a     a       1.2
## 3     2 b     b       2.1
## 4     3 c     c        13
```

For more options affecting how input files are interpreted, read the function documentation: ?read_csv.

Column Types

When read_csv parses a file, it infers the type of each column. This inference can be slow, or worse the inference can be incorrect. If you know a priori what the types should be, you can specify this using the col_types option. If you do this, then read_csv will not make a guess at the types. It will, however, replace values that it cannot parse as of the right type into NA.[2]

String-Based Column Type Specification

In the simplest string specification format, you must provide a string with the same length as you have columns and where each character in the string specifies the type of one column. The characters specifying different types are this:

Character	Type
c	Character
i	Integer
n	Number
d	Double
l	Logical
f	Factor
D	Date
T	Datetime
t	Time
?	Guess (default)
_/-	Skip the column

---

[2] There is a gotcha here. The types are guessed at after a fixed number of lines are read (by default 1000). If you have 1000 lines of numbers in a column and line 1001 has a string, then the type will be inferred as numeric and you lose the string. If you know the types, it is always better to tell the functions what they are.

By default, read_csv guesses, so we could make this explicit using the type specification "????":

```
read_csv(
    file ="data/data.csv",
    col_types ="????"
)
```

```
## # A tibble: 3 × 4
##        A B     C         D
##    <dbl> <chr> <chr> <dbl>
## 1      1 a     a       1.2
## 2      2 b     b       2.1
## 3      3 c     c        13
```

The results of the guesses are double for columns A and D and character for columns B and C. If we wanted to make *this* explicit, we could use "dccd".

```
read_csv(
    file ="data/data.csv",
    col_types ="dccd"
)
```

```
## # A tibble: 3 × 4
##        A B     C         D
##    <dbl> <chr> <chr> <dbl>
## 1      1 a     a       1.2
## 2      2 b     b       2.1
## 3      3 c     c        13
```

If you want an integer type for column A, you can use "iccd":

```
read_csv(
    file ="data/data.csv",
    col_types ="iccd"
)
```

```
## # A tibble: 3 × 4
##        A B     C         D
```

```
##    <int> <chr> <chr> <dbl>
## 1     1 a     a       1.2
## 2     2 b     b       2.1
## 3     3 c     c        13
```

If you try to interpret column D as integers as well, you will get a list of warning messages, and the values in column D will all be NA; the numbers in column D cannot be interpreted as integers, and read_csv will not round them to integers.

```
read_csv(
    file ="data/data.csv",
    col_types ="icci"
)
```

```
## Warning: One or more parsing issues, see
## 'problems()' for details
```

```
## # A tibble: 3 × 4
##       A B     C         D
##    <int> <chr> <chr> <int>
## 1     1 a     a        NA
## 2     2 b     b        NA
## 3     3 c     c        NA
```

If you specify that a column should have type d, the numbers in the column must be integers or decimal numbers. If you use the type n (the default that read_csv will guess), you will also get doubles, but the latter type can handle strings that can be interpreted as numbers such as dollar amounts, percentages, and group separators in numbers. The column type n will ignore leading and trailing text and handle number separators:

With this function call

```
read_csv(
    'A, B, C,    D,    E
    $1,a,a,1.2%,"1,100,200"
    $2,b,b,2.1%,"140,000"
    $3,c,c,13.0%,"2,005,000"',
    col_types ="nccnn"
    )
```

```
## # A tibble: 3 × 5
##       A B      C          D         E
##    <dbl> <chr> <chr>  <dbl>    <dbl>
## 1     1 a      a        1.2  1100200
## 2     2 b      b        2.1   140000
## 3     3 c      c         13  2005000
```

columns A, D, and E will be read as numbers. If you use the type specification d, they would not, and all the values would be NA.

The decimal indicator and group delimiter vary around the world. By default, read_csv uses the US convention with a dot for decimal notation and comma for grouping in numbers. In many European countries, it is the opposite. You can use the locale option to change these:

```
read_csv(
    'A, B, C,   D,    E
    $1,a,a," 1,2%","1.100.200"
    $2,b,b," 2,1%","   140.000"
    $3,c,c,"13,0%","2.005.000"',
    locale=locale(decimal_mark =",",grouping_mark ="."),
    col_types ="nccnn")
```

```
## # A tibble: 3 × 5
##       A B      C          D         E
##    <dbl> <chr> <chr>  <dbl>    <dbl>
## 1     1 a      a        1.2  1100200
## 2     2 b      b        2.1   140000
## 3     3 c      c         13  2005000
```

In the preceding example, I explicitly specified how read_csv should interpret numbers, but you can also use ISO 639-1 language codes.[3] If you do, you also get the local time conventions and local day and month names. The default is English, but if

---

[3] https://en.wikipedia.org/wiki/List_of_ISO_639-1_codes

your data is from Denmark, for example, you want to use Danish conventions, you would use the local locale("da"). For French data, you would use fr, locale("fr"). If you type this into an R console, you will see the month and week names, including their abbreviated forms, in these languages.

See the ?locale documentation for more options.

In files that use commas as decimal points and "." for number groupings, the column delimiter is usually ";" rather than ",". This way, it is not necessary to put decimal numbers in quotes. The read_csv2 function works as read_csv but uses ";" as column delimiter and "." for number groupings.

The *logical* type is used for boolean values. If a column only contains TRUE and FALSE (case doesn't matter)

```
read_csv(
    'A, B, C,    D
    TRUE, a, a,    1.2
    false, b, b,    2.1
    true, c, c, 13',
    show_col_types =FALSE
)
```

```
## # A tibble: 3 × 4
##   A      B     C     D
##   <lgl> <chr> <chr> <dbl>
## 1 TRUE  a     a      1.2
## 2 FALSE b     b      2.1
## 3 TRUE  c     c     13
```

then read_csv will guess that the type is logical.

It is not unusual to code boolean values as 0 and 1, however, and since these will be interpreted as numbers by default, you can make their type explicit using l:

```
read_csv(
    'A, B, C,    D
    1, a, a,    1.2
    0, b, b,    2.1
    1, c, c, 13',
    col_types ="lccn"
)
```

```
## # A tibble: 3 × 4
##   A     B     C         D
##   <lgl> <chr> <chr> <dbl>
## 1 TRUE  a     a       1.2
## 2 FALSE b     b       2.1
## 3 TRUE  c     c        13
```

If you use type l, you can mix TRUE/FALSE (ignoring case) with 0/1. Any other number or string will be translated into NA.

The D, t, and T types are for *dates*, *time points*, and *datetime*, in that order. Dates and time are what you might expect. A date specifies a range of days, for example, a single day, a week, a month, or a year. A time point specifies a specific time of the day, for example, an hour, a minute, or a second. A datetime combines a day and a time, that is, it specifies a specific time during a specific day.

```
read_csv(
    'D, T, t
    "2018-08-23", "2018-08-23T14:30", 14:30',
    col_types ="DTt"
)
```

```
## # A tibble: 1 × 3
##   D          T                   t
##   <date>     <dttm>              <time>
## 1 2018-08-23 2018-08-23 14:30:00 14:30
```

If you use one of these type specifications, the time and dates should be in ISO 8601 format.[4] Local conventions for writing time and date, however, differ substantially and are rarely ISO 8601. When your time data are not ISO 8601, you need to tell read_csv how to read them.

---

[4]https://en.wikipedia.org/wiki/ISO_8601

The default time parser handles times in the hh:mm, hh:mm:ss formats and handles am and pm suffixes; it suffices for most time formats (but notice that it wants time in hh:mm or hh:mm:ss format; it is flexible in the number of characters you use for hours, and you can leave out seconds, but you cannot leave out minutes). Date and datetime vary much more than time formats, and there, you usually need to specify the encoding format.

You can use the locale option to change how read_csv parses dates (D) and time (t).

```
read_csv(
    'D, t
    "23 Oct 2018", 2pm',
    col_types ="Dt",
    locale =locale(
        date_format ="%d %b %Y",
        time_format ="%I%p"
    )
)

## # A tibble: 1 × 2
##   D          t
##   <date>     <time>
## 1 2018-10-23 14:00
```

The date_format "%d %b %Y" says that dates are written as day, three-letter month abbreviation, and year with four digits, and each of the three separated by a space. The time_format "%I%p" says that we want time to be written as a number from 1 to 12, with no minute information, the hour immediately followed by am/pm without any space between.

For datetimes (T), we cannot specify the format using locale. We need a more verbose type specification that we will return to later. We also return to formatting specifications for parsing dates and time later.

Columns that are not immediately parsed as numbers, booleans, dates, or times will be parsed as *strings*. If you want these to be factors instead, you use the f type specification.

```
read_csv(
    'A, B, C,   D
    1, a, a,    1.2
    0, b, b,    2.1
    1, c, c,    13',
```

```
col_types ="lcfn")
```

```
## # A tibble: 3 × 4
##    A     B     C     D
##    <lgl> <chr> <fct> <dbl>
## 1 TRUE  a     a       1.2
## 2 FALSE b     b       2.1
## 3 TRUE  c     c        13
```

If you only want to use some of the columns, you can skip the rest using the "type" - or _:

```
read_csv(
    file ="data/data.csv",
    col_type ="_cc-"
)
```

```
## # A tibble: 3 × 2
##    B     C
##    <chr> <chr>
## 1 a     a
## 2 b     b
## 3 c     c
```

If you specify the column types using a string, you should specify the types of all columns. If you only want to define the types of a subset of columns, you can use the function cols() to specify types. You call this function with named parameters, where the names are column names and the arguments are types.

```
read_csv(
    file ="data/data.csv",
    col_types =cols(A ="c")
)
```

```
## # A tibble: 3 × 4
##    A     B     C     D
##    <chr> <chr> <chr> <dbl>
## 1 1     a     a       1.2
## 2 2     b     b       2.1
## 3 3     c     c        13
```

```
read_csv(
    file ="data/data.csv",
    col_types =cols(A ="c",D ="c")
)

## # A tibble: 3 × 4
##    A     B     C     D
##    <chr> <chr> <chr> <chr>
## 1 1     a     a     1.2
## 2 2     b     b     2.1
## 3 3     c     c     13.0
```

Function-Based Column Type Specification

If you are like me, you might find it hard to remember the single-character codes for different types. If so, you can use longer type names that you specify using function calls. These functions have names that start with col_, so you can use autocomplete to get a list of them. The types you can specify using functions are the same as those you can specify using characters, of course, and the functions are as follows:

Function	Type
col_character()	Character
col_integer()	Integer
col_number()	Number
col_double()	Double
col_logical()	Logical
col_factor()	Factor
col_date()	Date
col_datetime()	Datetime
col_time()	Time
col_guess()	Guess (default)
col_skip()	Skip the column

You need to wrap the function-based type specifications in a call to cols.

```
read_csv(
    file ="data/data.csv",
    col_types =cols(A =col_integer())
)
```

```
## # A tibble: 3 × 4
##       A B     C         D
##    <int> <chr> <chr> <dbl>
## 1     1 a     a       1.2
## 2     2 b     b       2.1
## 3     3 c     c        13
```

```
read_csv(
    file ="data/data.csv",
    col_types =cols(D =col_character())
)
```

```
## # A tibble: 3 × 4
##       A B     C     D
##    <dbl> <chr> <chr> <chr>
## 1     1 a     a     1.2
## 2     2 b     b     2.1
## 3     3 c     c     13.0
```

Most of the col_ functions do not take any arguments, but they are affected by the locale parameter the same way that the string specifications are.

For factors, date, time, and datetime types, however, you have more control over the format using the col_ functions. You can use arguments to these functions for specifying how read_csv should parse dates and how it should construct factors.

For factors, you can explicitly set the levels. If you do not, then the column parser will set the levels in the order it sees the different strings in the column. For example, in data/data.csv the strings in columns C and D are in the order a, b, and c:

```
A, B, C,   D
1, a, a,   1.2
2, b, b,   2.1
3, c, c, 13.0
```

By default, the two columns will be interpreted as characters, but if we specify that C should be a factor, we get one where the levels are a, b, and c, in that order.

```
my_data<-read_csv(
    file = "data/data.csv",
    col_types = cols(C =col_factor())
)
my_data$C
```

```
## [1] a b c
## Levels: a b c
```

If we want the levels in a different order, we can give col_factor() a levels argument.

```
my_data<-read_csv(
    "A, B, C
      Foo, 12.4, Medium
      Bar, 5.2,   Small
      Baz, 42.0, Large
      ",
    col_types =cols(
        C =col_factor(levels =c("Small","Medium","Large"))
    )
)
my_data$C
```

```
## [1] Medium Small   Large
## Levels: Small Medium Large
```

We can also make factors ordered using the ordered argument.

```
my_data<-read_csv(
    file ="data/data.csv",
    col_types =cols(
        B =col_factor(ordered =TRUE),
        C =col_factor(levels =c("c","b","a"))
    )
)
```

```
my_data$B
```

```
## [1] a b c
## Levels: a b c
```

```
my_data$C
```

```
## [1] a b c
## Levels: c b a
```

Parsing Time and Dates

The most complex types to read (or write) are dates *and* time (and datetime), just because these are written in many different ways. You can specify the format that dates and datetime are in using a string with codes that indicate how time information is represented.

The codes are these:

Code	Time format	Example string	Interpretation
%Y	4-digit year	1975	The year 1975
%y	2-digit year[5]	75	Also the year 1975
%m	2-digit month	02	February
%b	Abbreviated month name[6]	Feb	February
%B	Full month name	February	February
%d	2-digit day	15	The 15th of a month
%H	Hour number on a 24-hour clock	18	Six o'clock in the evening

(continued)

---

[5] Two-digit years are assumed to be either in the 20th or the 21st century. The cutoff line is 68; years at or below 68 are in the 20th century, and 69 and above are in the 21st century. Therefore, 75 is assumed to be in the 20th century, so 75 is 1975.

[6] The name of months and weekdays varies from language to language, and so does the abbreviations. Therefore, if you use a format that refers to these, you either need to use numbers or the format will depend on the locale option.

Code	Time format	Example string	Interpretation
%I	Hour number on a 12-hour clock[7]	6 pm	18:00 hours
%p	AM/PM indicator	6 pm	18:00 hours
%M	Two-digit minutes	18:30	Half past six
%S	Integer seconds	18:30:10	Ten seconds past 18:00
%Z	Time zone as name[8]	America/Chicago	Central Time
%z	Time zone as offset from UTC	"+0100"	Central European Time

There are shortcuts for frequently used formats:

Shortcut	Format
%D	%m/%d/%y
%x	%y/%m/%d
%F	%Y-%m-%d
%R	%H:%M
%T	%H:%M:%S

As we saw earlier, you can set the date and time format using the locale() function. If you do not, the default codes will be %AD for dates and %AT for time (there is no locale() argument for datetime). These codes specify YMD and H:M/H:M:S formats, respectively, but are more relaxed in matching the patterns. The date parse, for example, will allow different separators. For dates, both "1975-02-15" and "1975/02/15" will be read as February 15, 1975, and for time, both "18:00" and "6:00 pm" will be six o'clock in the evening.

---

[7] If you use %I, you must also use %p for pm/am.

[8] Time zones based on locations handle daylight saving time. UTC is a fixed time zone; it does not have daylight saving time.

In the following text, I give a few examples. I will use the functions `parse_date`, `parse_time`, and `parse_datetime` rather than `read_csv` with column type specifications. These functions are used by `read_csv` when you specify a date, time, or datetime column type, but using `read_csv` for the examples would be unnecessarily verbose. Each takes a vector string representation of dates and time. For more examples, you can read the function documentation `?col_datetime`.

Parsing time is simplest; there is not much variation in how time points are written. The main differences are in whether you use 24-hour clocks or 12-hour clocks. The %R and %T codes expect 24-hour clocks and differ in whether seconds are included or not.

```
parse_time(c("18:00"),format ="%R")
```

```
## 18:00:00
```

```
parse_time(c("18:00:30"),format ="%T")
```

```
## 18:00:30
```

There is no shortcut for 12-hour codes, but you must combine %I with %p to read PM/AM formats.

```
parse_time(c("6 pm"),format ="%I %p")
```

```
## 18:00:00
```

Here, I have specified that the input only includes hours and not minutes. If we want hours (and not minutes) in 24-hour clocks, we need to use %H rather than %R.

```
parse_time(c("18"),format ="%R")
```

```
## 18:00:00
```

For dates, ISO 8601 says that the format should be YYYY-MM-DD. The default date parser will accept this format, but the explicit format string is

```
parse_date(c("1975-02-05"),format ="%Y-%m-%d")
```

```
## [1] "1975-02-05"
```

If you do not want to include the day, and you want to use two-digit years, you need

```
parse_date(c("75-02"),format ="%y-%m")
```

```
## [1] "1975-02-01"
```

26

This is February 1975; remember that the %y code assumes that numbers above 68 are in the 20th century.

Dates written on the form 15/02/75 can mean both February 15, 1975, and May 2, 1975, depending on where you are in the world. Europe uses the sensible DD/MM/YY format, where the order goes from the smallest time unit, days, to the medium time units, months, and then to years. In the United States, they use the MM/DD/YY format. To get the 15th of February, you need one of these formats:

```
parse_date(c("15/02/75"),format ="%d/%m/%y")
```

```
## [1] "1975-02-15"
```

```
parse_date(c("02/15/75"),format ="%m/%d/%y")
```

```
## [1] "1975-02-15"
```

Date specifications that only use numbers are not affected by the local language, but if you include the name of months, they are. The name of months and their abbreviation varies from language to language, obviously. So does the name of weekdays, but at the time of writing, parsing weeks and weekdays is not supported by readr. You can get the name information from locale() if you use a language code. In the following examples, I parse dates in English and Danish. The month names are almost the same, but abbreviations in Danish require a dot following them, and the day is followed by a dot as well.

```
parse_date(c("Feb 15 1975"),format ="%b %d %Y",locale =locale("en"))
```

```
## [1] "1975-02-15"
```

```
parse_date(c("15. feb. 1975"),format ="%d. %b %Y",locale =locale("da"))
```

```
## [1] "1975-02-15"
```

```
parse_date(c("February 15 1975"),format ="%B %d %Y",locale =locale("en"))
```

```
## [1] "1975-02-15"
```

```
parse_date(c("15. feb. 1975"),format ="%d. %b %Y",locale =locale("da"))
```

```
## [1] "1975-02-15"
```

```
parse_date(c("Oct 15 1975"),format ="%b %d %Y",locale =locale("en"))
```

```
## [1] "1975-10-15"

parse_date(c("15. okt. 1975"),format ="%d. %b %Y",locale =locale("da"))

## [1] "1975-10-15"

parse_date(c("October 15 1975"),format ="%B %d %Y",locale =locale("en"))

## [1] "1975-10-15"

parse_date(c("15. oktober 1975"),format ="%d. %B %Y",locale =locale("da"))

## [1] "1975-10-15"
```

"Datetimes" can be parsed using combinations of date and time strings. With these, you also want to consider time zones. You can ignore those for dates and time, but unless you are sure that you will never have to consider time zones, you should not rely on the default time zone (which is UTC).[9]

You can either specify that time zones are relative to UTC with %z or location based, with %Z if the time zone is given in the input, or you can use locale() if it is the same for all the input.

If you specify a time zone based on a location, R will automatically adjust for daylight saving time, but if you use dates relative to UTC, you will not—UTC does not have daylight savings. Central European Time (CET) is "+0100" and with daylight saving time "+0200". US Pacific Time (PST) is "-0800", but with daylight saving time (PDT), it is "-0700". When you switch back and forth between daylight savings is determined by your location.

These two datetimes are the same:

```
parse_datetime(c("Feb 15 1975 18:00 US/Pacific"),format ="%b %d %Y %R %Z")

## [1] "1975-02-16 02:00:00 UTC"

parse_datetime(c("Feb 15 1975 18:00 -0800"),format ="%b %d %Y %R %z")

## [1] "1975-02-16 02:00:00 UTC"
```

---

[9]https://en.wikipedia.org/wiki/Coordinated_Universal_Time

as are these two:

```
parse_datetime(c("Jun 15 1975 18:00 US/Pacific"),format ="%b %d %Y %R %Z")
```

```
## [1] "1975-06-16 01:00:00 UTC"
```

```
parse_datetime(c("Jun 15 1975 18:00 -0700"),format ="%b %d %Y %R %z")
```

```
## [1] "1975-06-16 01:00:00 UTC"
```

If you use locale() to specify a time zone, you cannot use zones relative to UTC. The point of using locale() is local formats, not time zones. The parser will still handle daylight savings for you, however. These two are the same datetimes:

```
parse_datetime(
    c("Aug 15 1975 18:00"),
    format ="%b %d %Y %R",
    locale =locale(tz ="US/Pacific")
)
```

```
## [1] "1975-08-15 18:00:00 PDT"
```

```
parse_datetime(
    c("Aug 15 1975 18:00 US/Pacific"),
    format ="%b %d %Y %R %Z"
)
```

```
## [1] "1975-08-16 01:00:00 UTC"
```

They are printed differently, but as we saw earlier, 6 pm (18:00) PDT is the same as 01:00 (the following day) in UTC.

If you print the objects you parse, there is a difference between using locale() and using %Z, but the time will be the same. Using %Z, you will automatically translate the time into UTC; using locale(), you will not. But you can compare the result of the two calls and see that they are equivalent objects:

```
x<-parse_datetime(
    c("Aug 15 1975 18:00"),
    format ="%b %d %Y %R",
    locale =locale(tz ="US/Pacific")
)
```

```
y<-parse_datetime(
    c("Aug 15 1975 18:00 US/Pacific"),
    format ="%b %d %Y %R %Z"
)
x==y
```

```
## [1] TRUE
```

The output of `parse_datetime()` looks like strings when you print them, but the object classes are not `character`, which, among other things, is why the comparison works.

Space-Separated Columns

The preceding functions all read delimiter-separated columns. They expect a single character to separate one column from the next. If the argument `trim_ws` is true, they ignore whitespace. This argument is true by default for `read_csv`, `read_csv2`, and `read_tsv`, but false for `read_delim`.

The function `read_table` instead separates columns by one or more spaces:

```
read_table(
    "A    B    C    D
     1    2    3    4
     15   16   17   18"
)
```

```
## # A tibble: 2 × 4
##        A      B      C      D
##    <dbl>  <dbl>  <dbl>  <dbl>
## 1      1      2      3      4
## 2     15     16     17     18
```

The first line is interpreted as a header, just as for the previous functions, but you can disable that with `col_names = FALSE` again:

```
read_table(
    "A    B    C    D
     1    2    3    4
```

```
    15  16  17  18",
    col_names =FALSE
)

## # A tibble: 3 × 4
##    X1    X2    X3    X4
##    <chr> <chr> <chr> <chr>
## 1 A     B     C     D
## 2 1     2     3     4
## 3 15    16    17    18
```

The read_table function takes many of the same arguments as read_csv or read_tsv and mostly differs in what it considers the column separator—this function uses whitespace instead of a specific field separator such as a comma or a semicolon.

The package readxl is not loaded when you load the package tidyverse, but can be quite useful. Its read_excel function does exactly what it says on the tin; it reads Excel spreadsheets into R. Its interface is similar to the functions in readr. Where the interface differs is in Excel-specific options such as which sheet to read. Such options are clearly only needed when reading Excel files.

Functions for Writing Data

Writing data to a file is more straightforward than reading data because we have the data in the correct types and we do not need to deal with different formats. With readr's writing functions, we have fewer options to format our output—for example, we cannot give the functions a locale() and we cannot specify date and time formatting, but we can use different functions to specify delimiters and time will be output in ISO 8601 which is what the reading functions will use as default.

The functions are write_delim, write_csv, write_csv2, and write_tsv, and for formats that Excel can read, write_excel_csv and write_excel_csv2. The difference between write_csv and write_excel_csv and between write_csv2 and write_excel_csv2 is that the Excel functions include a UTF-8 byte order mark so Excel knows that the file is UTF-8 encoded.

The first argument to these functions is the data we want to write, and the second is the path to the file we want to write to. If this file has suffix .gz, .bz2, or .xz, the output is automatically compressed.

I will not list all the arguments for these functions here, but you can read the documentation for them from the R console. The argument you are most likely to use is col_names, which, if true, means that the function will write the column names as the first line in the output, and if false, it will not. If you use write_delim, you might also want to specify the delimiter character using the delim argument. By default, it is a single space; if you write to a file using write_delim with the default options, you get the data in a format that you can read using read_table.

The delimiter characters and the decimal points for write_csv, write_csv2, and write_tsv are the same as for the corresponding read functions.

## CHAPTER 3

Representing Tables: tibble

The data that the readr package returns are represented as tibble objects. These are tabular data representations similar to the base R data frames but are a more modern version.

The package that implements tibbles is tibble. You can load it using

```
library(tibble)
```

or as part of the tidyverse:

```
library(tidyverse)
```

Creating Tibbles

Tidyverse functions that create tabular data will create tibbles rather than data frames. For example, when we use read_csv to read a file into memory, the result is a tibble:

```
x<-read_csv(file ="data/data.csv",show_col_types =FALSE)
x

## # A tibble: 3 × 4
##       A B     C           D
##   <dbl> <chr> <chr>   <dbl>
## 1     1 a     a         1.2
## 2     2 b     b         2.1
## 3     3 c     c          13
```

© Thomas Mailund 2022
T. Mailund, *R 4 Data Science Quick Reference*, https://doi.org/10.1007/978-1-4842-8780-4_3

The table that `read_csv()` creates has several superclasses, but the last is `data.frame`.

```
class(x)
```

```
## [1] "spec_tbl_df" "tbl_df"     "tbl"
## [4] "data.frame"
```

This means that generic functions, if not specialized in the other classes, will use the `data.frame` version, and this, in turn, means that you can often use tibbles in functions that expect data frames. It does not mean that you can *always* use tibbles as a replacement for a data frame. If you run into this problem, you can translate a tibble into a data frame using `as.data.frame()`:

```
y<-as.data.frame(x)
y
```

```
##   A B C    D
## 1 1 a a  1.2
## 2 2 b b  2.1
## 3 3 c c 13.0
```

```
class(y)
```

```
## [1] "data.frame"
```

Notice that the two objects, when printed, give different output. When printing a tibble, there is more information about the column types.

If you have a data frame, you can translate it into a tibble using `as_tibble()`:

```
z<-as_tibble(y)
z
```

```
## # A tibble: 3 × 4
##       A B     C         D
##   <dbl> <chr> <chr> <dbl>
## 1     1 a     a       1.2
## 2     2 b     b       2.1
## 3     3 c     c      13
```

You can create a tibble from vectors using the tibble() function:

```
x <- tibble(
    x = 1:100,
    y = x^2,
    z = y^2
)
x
```

```
## # A tibble: 100 × 3
##          x      y      z
##      <int> <dbl>  <dbl>
## 1       1      1      1
## 2       2      4     16
## 3       3      9     81
## 4       4     16    256
## 5       5     25    625
## 6       6     36   1296
## 7       7     49   2401
## 8       8     64   4096
## 9       9     81   6561
## 10     10    100  10000
## # . . . with 90 more rows
```

Two things to notice here: when you print a tibble, you only see the first 10 lines. This is because the tibble has enough lines that it will flood the console if you print all of them. If a tibble has more than 20 rows, you will only see the first 10. If it has fewer, you will see all the rows.

You can change how many lines you will see using the n option to print():

```
print(x,n =2)
```

```
## # A tibble:  100 × 3
##          x      y      z
##      <int> <dbl> <dbl>
## 1       1      1      1
## 2       2      4     16
## # . . . with 98 more rows
```

If a tibble has more columns than your console can show, only some will be printed. You can change the number of characters it will print using the width option to print.

```
print(x,n =2,width =15)
```

```
## # A tibble:
## #   100 × 3
##       x     y
##    <int> <dbl>
## 1     1     1
## 2     2     4
## # . . . with 98
## #   more rows,
## #   and 1 more
## #   variable:
## #   z <dbl>
```

You can set either option to Inf. If n is Inf, you will see all rows, and if width is Inf, you will see all the columns.

The second thing to notice is that you can refer to previous columns when specifying later columns. When we created x, we used

```
tibble(
    x =1:100,
    y =x^2,
    z =y^2
)
```

where column y refers to column x and column z refers to column y. You cannot refer to a variable in the following columns, so this would be an error:

```
tibble(
    w =x/2,
    x =1:100
)
```

When you use tibble() to create a data frame, you specify the columns as named arguments. If you indent your code as I have earlier, then you can think of this as defining each column in one line. You can also create a tibble with one line per row.

```
tribble(
    ~x,~y, ~z,
     1,10,100,
     2,20,200,
     3,30,300
)
```

```
## # A tibble: 3 × 3
##       x     y     z
##   <dbl> <dbl> <dbl>
## 1     1    10   100
## 2     2    20   200
## 3     3    30   300
```

The first line names the columns, and the ~ is necessary before the names. For large tibbles, using `tribble()` is not that helpful, but for example code or small tables, it can be.

Indexing Tibbles

You can index a tibble in much the same way as you can index a data frame. You can extract a column using a single-bracket index ([ ]), either by name or by index:

```
x<-read_csv(file ="data/data.csv",show_col_types =FALSE)
y<-as.data.frame(x)
x["A"]
```

```
## # A tibble: 3 × 1
##       A
##   <dbl>
## 1     1
## 2     2
## 3     3
```

```
y["A"]
```

```
##   A
## 1 1
```

```
## 2 2
## 3 3
```

```
x[1]
```

```
## # A tibble: 3 × 1
##       A
##   <dbl>
## 1     1
## 2     2
## 3     3
```

```
y[1]
```

```
##   A
## 1 1
## 2 2
## 3 3
```

The result is a `tibble` or `data.frame`, respectively, containing a single column.

If you use double brackets (`[[]]`), you will get the vector contained in a column rather than a tibble/data frame:

```
x[["A"]]
```

```
## [1] 1 2 3
```

```
y[["A"]]
```

```
## [1] 1 2 3
```

You will also get the underlying vector of a column if you use $ indexing:

```
x$A
```

```
## [1] 1 2 3
```

```
y$A
```

```
## [1] 1 2 3
```

Using [ ], you can extract more than one column.

```
x[c("A","C")]
```

```
## # A tibble: 3 × 2
##       A C
```

```
##    <dbl> <chr>
## 1     1 a
## 2     2 b
## 3     3 c
```

```
y[c("A","C")]
```

```
##   A C
## 1 1 a
## 2 2 b
## 3 3 c
```

```
x[1:2]
## # A tibble: 3 × 2
##       A B
##    <dbl> <chr>
## 1     1 a
## 2     2 b
## 3     3 c
```

```
y[1:2]
```

```
##   A B
## 1 1 a
## 2 2 b
## 3 3 c
```

You cannot do this using [[ ]].

You can extract a subset of rows and columns if you use two indices. For example, you can get the first two rows in the first two columns using [1:2,1:2]:

```
x[1:2,1:2]
```

```
## # A tibble: 2 × 2
```

```
##          A B
##      <dbl> <chr>
## 1        1 a
## 2        2 b
```

```
y[1:2,1:2]
```

```
##    A B
## 1 1 a
## 2 2 b
```

With a single index, you always extract a subset of columns. If you want to extract a subset of rows for all columns, you can use

```
x[1:2,]
```

```
## # A tibble: 2 × 4
##          A B     C         D
##      <dbl> <chr> <chr> <dbl>
## 1        1 a     a       1.2
## 2        2 b     b       2.1
```

```
y[1:2,]
```

```
##    A B C   D
## 1 1 a a 1.2
## 2 2 b b 2.1
```

If you extract a subset of rows from a single column, tibbles and data frames no longer have the same behavior. A tibble will give you a tibble in return, while a data frame will give you a vector:

```
x[1:2,2]
```

```
## # A tibble: 2 × 1
##    B
##    <chr>
## 1 a
```

```
## 2 b
```

```
y[1:2,2]
```

```
## [1] "a" "b"
```

Tibbles are more consistent than `data.frame` objects. When you extract part of a tibble, you always get a tibble in return. Data frames sometimes give you a data frame and sometimes (in this particular case) a vector. In any function that expects a data frame and only use subscripts that return a data frame, you can also use a tibble. Tibbles return tibbles, and since a tibble is also a data frame, this will not cause any problems. If a function expects to extract a vector, you cannot use a tibble. This is where you will need to use `as.data.frame()` to get a `data.frame()` with data frame behavior.

CHAPTER 4

Tidy Select

So-called "tidy select" is not a package you would use on its own (although you can import it and it is called `tidyselect`), rather it is a small language for selecting columns in a data frame or tibble. Many packages in the following chapters support this language, so rather than describing it in each chapter, I decided to put a description in its own chapter, and this is as good a place as any. We cannot, however, use the language without functionality from packages that are described in later chapters, so I will use the function `select` from the `dplyr` package; see Chapter 8.

The select function helps you select columns from a tibble or a data frame. Its first argument is a table, so we can use it with a pipe operator, and then the remaining can be the columns we want to extract.

```
tbl<-tribble(
  ~sample,~min_size,~max_size,~min_weight,~max_weight,
  "foo",13,16,45.2,67.2,
  "bar",12,17,83.1,102.5
)
tbl|>select(sample, min_size, min_weight)

## # A tibble: 2 × 3
## sample min_size min_weight
##   <chr>    <dbl>      <dbl>
## 1 foo         13       45.2
## 2 bar         12       83.1
```

This simple way of selecting columns is useful in itself, of course, but doesn't add much on top of just indexing into tables. The power of tidy select is the small language that we can use instead of explicitly listing columns.

© Thomas Mailund 2022
T. Mailund, *R 4 Data Science Quick Reference*, https://doi.org/10.1007/978-1-4842-8780-4_4

Ranges

If you want a range of columns, say from min_size to max_weight, you can use : to select the columns:

```
tbl|>select(min_size:max_weight)
```

```
## # A tibble: 2 × 4
## min_size max_size min_weight max_weight
##    <dbl>    <dbl>      <dbl>      <dbl>
## 1     13       16       45.2       67.2
## 2     12       17       83.1      102.
```

This only works for contiguous columns, but you can select multiple ranges if you want:

```
tbl|>select(sample:min_size, min_weight:max_weight)
```

```
## # A tibble: 2 × 4
##    sample min_size min_weight max_weight
##     <chr>    <dbl>      <dbl>      <dbl>
## 1  foo        13       45.2       67.2
## 2  bar        12       83.1      102.
```

Complements

Sometimes, it is easier to specify which columns you do *not* want, and then you can use the complement operator !. This works both for the complement of a single column, so "everything except this column":

```
tbl|>select(!sample)
```

```
## # A tibble: 2 × 4
## min_size max_size min_weight max_weight
##    <dbl>    <dbl>      <dbl>      <dbl>
## 1     13       16       45.2       67.2
## 2     12       17       83.1      102.
```

but it also works if you select columns some other way, for example, through ranges. If sample:min_size selects the columns sample and min_size, then !sample:min_size is everything except those columns:

```
tbl|>select(!(sample:min_size))
```

```
## # A tibble: 2 × 3
##   max_size min_weight max_weight
##      <dbl>      <dbl>      <dbl>
## 1       16       45.2       67.2
## 2       17       83.1      102.
```

Unions and Intersections

It gets a bit strange if you include both the complements of columns and the columns themselves, but you will get all the columns you ask for, so with

```
tbl|>select(sample,!(sample:min_size))
```

```
## # A tibble: 2 × 4
## sample max_size min_weight max_weight
##    <chr>    <dbl>      <dbl>      <dbl>
## 1 foo         16       45.2       67.2
## 2 bar         17       83.1      102.
```

column sample, and then we get all the columns that are not in the range "sample:min_size".

You can explicitly ask for unions or intersections of selections using & (intersection) and | (union).

```
tbl|>select(
  sample:min_weight   # sample, min_size, max_size, and min_weight
  &                   # intersect with
  max_size:max_weight # max_size, min_weight, max_weight
)
```

```
## # A tibble: 2 × 2
##    max_size min_weight
##       <dbl>      <dbl>
## 1        16       45.2
## 2        17       83.1
```

```
tbl|>select(
    sample:min_weight      # sample, min_size, max_size, and min_weight
    |                      # union with
    max_size:max_weight    # max_size, min_weight, max_weight
)
```

```
## # A tibble: 2 × 5
##    sample min_size max_size min_weight max_weight
##    <chr>     <dbl>    <dbl>      <dbl>      <dbl>
## 1 foo          13       16       45.2       67.2
## 2 bar          12       17       83.1       102.
```

Select Columns Based on Name

There are several functions that will let you select columns based on their names. The starts_with() function will pick columns that start with a specific string:

```
tbl|>select(starts_with("min"))
```

```
## # A tibble: 2 × 2
##    min_size min_weight
##       <dbl>      <dbl>
## 1        13       45.2
## 2        12       83.1
```

Similarly, the ends_with() function will let you pick columns with names that end with a specific string:

```
tbl|>select(ends_with("weight"))
```

```
## # A tibble: 2 × 2
##    min_weight max_weight
```

```
##          <dbl>       <dbl>
## 1        45.2        67.2
## 2        83.1       102.
```

If you want to pick all columns whose names contain a string, you use the function contains():

```
tbl|>select(contains("_"))
```

```
## # A tibble: 2 × 4
## min_size max_size min_weight max_weight
##    <dbl>    <dbl>      <dbl>      <dbl>
## 1     13       16       45.2       67.2
## 2     12       17       83.1      102.
```

For more complex matching, the matches() function lets you pick columns based on regular expressions. Here, I'll just pick the columns with names that contain an underscore gain, because there really isn't anything complicated to select for in this example table:

```
tbl|>select(matches(".*_.*"))
```

```
## # A tibble: 2 × 4
## min_size max_size min_weight max_weight
##    <dbl>    <dbl>      <dbl>      <dbl>
## 1     13       16       45.2       67.2
## 2     12       17       83.1      102.
```

Everything

It might not seem like much of a feature, since selecting all columns just gives you the original table back, but there is a function, everything(), that does just that.

```
tbl|>select(everything())
```

```
## # A tibble: 2 × 5
##    sample min_size max_size min_weight max_weight
##    <chr>     <dbl>    <dbl>      <dbl>      <dbl>
## 1 foo          13       16       45.2       67.2
## 2 bar          12       17       83.1      102.
```

You wouldn't use that with select(), of course, because it doesn't do anything, but there are other functions that requires that you select columns for some transformation of the data, and there everything() can be used to pick all the columns in a data frame.

Indexing from the Last Column

If you just want to select the last column, you can use the function last_col():

```
tbl|>select(last_col())
```

```
## # A tibble: 2 × 1
##    max_weight
##         <dbl>
## 1        67.2
## 2       102.
```

but you can also use the function to select by indexing from the right. If you give it an integer argument, i, it will select the column that is i from the right:

```
tbl|>select(last_col(0))
```

```
## # A tibble: 2 × 1
##    max_weight
##         <dbl>
## 1        67.2
## 2       102.
```

```
tbl|>select(last_col(1))
```

```
## # A tibble: 2 × 1
##    min_weight
##         <dbl>
## 1        45.2
## 2        83.1
```

```
tbl|>select(last_col(2))
```

```
## # A tibble: 2 × 1
##     max_size
```

```
##        <dbl>
## 1         16
## 2         17
```

```
tbl|>select(last_col(3))
```

```
## # A tibble: 2 × 1
##   min_size
##      <dbl>
## 1       13
## 2       12
```

and you can, of course, use this in ranges:

```
tbl|>select(last_col(3):last_col())
```

```
## # A tibble: 2 × 4
##   min_size max_size min_weight max_weight
##      <dbl>    <dbl>      <dbl>      <dbl>
## 1       13       16       45.2       67.2
## 2       12       17       83.1      102.
```

Selecting from Strings

If you have the column names you want to select in a vector of strings

```
vars<-c("min_size","min_weight")
```

you can use the all_of() or any_of() functions to pick them:

```
tbl|>select(all_of(vars))
```

```
## # A tibble: 2 × 2
##   min_size min_weight
##      <dbl>      <dbl>
## 1       13       45.2
## 2       12       83.1
```

```
tbl|>select(any_of(vars))
```

```
## # A tibble: 2 × 2
##   min_size min_weight
##      <dbl>      <dbl>
## 1        13       45.2
## 2        12       83.1
```

The difference between the two functions is that all_of() considers it an error if vars contain a name that isn't found in the table, while any_of() does not.

```
vars<-c(vars,"foo")
tbl|>select(all_of(vars))
```

```
## Error in 'select()':
## ! Can't subset columns past the end.
## x Column 'foo' doesn't exist.
```

```
tbl|>select(any_of(vars))
```

```
## # A tibble: 2 × 2
##   min_size min_weight
##      <dbl>      <dbl>
## 1        13       45.2
## 2        12       83.1
```

Selecting Columns Based on Their Content

Perhaps the most powerful selection function is where. You give it a function as an argument, that function is called with each column, and the columns for which the function returns TRUE are selected.

So, for example, if you only want columns that are numeric, you can use where combined with is.numeric:

```
tbl|>select(where(is.numeric))
```

```
## # A tibble: 2 × 4
## min_size max_size min_weight max_weight
##    <dbl>    <dbl>      <dbl>      <dbl>
## 1       13       16       45.2       67.2
## 2       12       17       83.1      102.
```

Or if you want columns that are numeric and the largest value is greater than 100, you can use

```
tbl|>select(where(\(x)is.numeric(x)&&max(x)>100.0))
```

```
## # A tibble: 2 × 1
##    max_weight
##         <dbl>
## 1        67.2
## 2       102.
```

(The syntax \(args) body was introduced in R 4 as a short form of function(args) body but means the same thing).

It Is a Growing Language, so Check for Changes

The language used by functions that support tidy select is evolving and growing, so there is likely more you can do by the time you read this. Check the documentation of the select function

```
?select
```

to get the most recent description.

CHAPTER 5

Reformatting Tables: tidyr

Even if we only consider tabular data, there are still many different ways to format this data. The packages in the Tidyverse expect that data is represented as so-called "tidy data"[1] (which is where the name Tidyverse comes from). The `tidyr` package helps you with formatting your data into tidy data.

You can load `tidyr` as part of the tidyverse:

```
library(tidyverse)
```

or on its own

```
library(tidyr)
```

Tidy Data

The fundamental properties that characterize tidy data are that each variable is in a column, and each observation is a row. The terms like variables and observations should be familiar from statistics, but what the tidy data properties say is that you should not put the values of a variable in different columns and you should not put more than one observation in the same row.

Consider data such as this:

```
mean_income<-tribble(
    ~country,~  '2001',~ '2002',~ '2003',~ '2004',~ '2005',
    "Numenor",  123456, 132654, 321646, 324156, 325416,
    "Westeros", 314256, 432165, 546123, 465321, 561423,
```

---

[1] https://en.wikipedia.org/wiki/Tidy_data

© Thomas Mailund 2022
T. Mailund, *R 4 Data Science Quick Reference*, https://doi.org/10.1007/978-1-4842-8780-4_5

```
    "Narnia",    432156,  342165,  564123,  543216,  465321,
    "Gondor",    531426,  321465,  235461,  463521,  561423,
    "Laputa",     14235,   34125,   45123,   51234,   54321
)
```

We could imagine that this table contains the mean income in five (fictional) countries in the first five years of the 21st century. Let us say that you are interested in knowing if there is a pattern in income over time when taking the different countries into account. To do this, you might consider a formula such as income ~ country + year, but the data is not formatted in a way that makes this easy. If year is a variable that you can use in the model, it is a problem that different observations, the actual years, are split into different columns. Similarly, if the mean income per year is a variable, it is a problem that all the observations for a country are in the same row. You cannot use income ~ country + year because the data is not tidy.

Pivoting

The function pivot_longer() can merge several columns into two—a key and a value column. The key column will be a discrete variable with values taken from the column names, and the value column will contain the data from the original columns.

Consider the mean income data frame. We want the actual years to be different outcomes of a year variable and have the income for each year to be another variable, income. We can get such a tibble by calling pivot_longer() with these parameters:

```
mean_income |> pivot_longer(
    names_to = "year",
    values_to = "income",
    cols = c(`2001`, `2001`, `2002`,
    `2003`, `2004`, `2005`)
)
```

```
## # A tibble: 25 × 3
##    country year  income
##    <chr>   <chr> <dbl>
## 1 Numenor  2001  123456
## 2 Numenor  2002  132654
```

```
## 3 Numenor  2003  321646
## 4 Numenor  2004  324156
## 5 Numenor  2005  325416
## 6 Westeros 2001  314256
## 7 Westeros 2002  432165
## 8 Westeros 2003  546123
## 9 Westeros 2004  465321
## 10 Westeros 2005  561423
## # . . . with 15 more rows
```

The first argument is the input data from mean_income, and I provide it using the |> operator. This works exactly as if I had given mean_income as the first argument to the function call, but using the operator signals to the reader that mean_income is data flowing into the function and different from other arguments because of it.

The names_to and values_to arguments are the names of two new columns. In the names_to column, we get the names of the columns we get the values from, and in the value column, we get the matching values from those columns. The remaining parameters specify which columns we wish to merge.

In the preceding example, I explicitly listed all the columns to include, but tidyr's functions support tidy select (see Chapter 4), so you can also use a range of columns, like this:

```
mean_income|>pivot_longer(
    names_to ="year",
    values_to ="income",
    cols = '2001':'2005'
)
## # A tibble: 25 × 3
##    country  year   income
##    <chr>    <chr>  <dbl>
## 1 Numenor  2001  123456
## 2 Numenor  2002  132654
## 3 Numenor  2003  321646
## 4 Numenor  2004  324156
## 5 Numenor  2005  325416
## 6 Westeros 2001  314256
## 7 Westeros 2002  432165
```

```
##  8 Westeros 2003   546123
##  9 Westeros 2004   465321
## 10 Westeros 2005   561423
## # . . . with 15 more rows
```

You can use more than one range:

```
mean_income|>pivot_longer(
    names_to ="year",
    values_to ="income",
    cols =c( '2001':'2002', '2004':'2005')
)
```

but the range specification is only useful when the relevant columns are contiguous. You can also specify the complement of the columns you want to merge using !. For example, if you want to merge the years, you can pick all columns except for country rather than pick all the years:

```
mean_income|>pivot_longer(
    names_to ="year",
    values_to ="income",
    cols =!country
)
```

```
## # A tibble: 25 × 3
##    country year  income
##    <chr>   <chr> <dbl>
##  1 Numenor 2001  123456
##  2 Numenor 2002  132654
##  3 Numenor 2003  321646
##  4 Numenor 2004  324156
##  5 Numenor 2005  325416
##  6 Westeros 2001 314256
##  7 Westeros 2002 432165
##  8 Westeros 2003 546123
##  9 Westeros 2004 465321
## 10 Westeros 2005 561423
## # . . . with 15 more rows
```

You can specify more than one column in a complement, and you can combine complements with ranges.

Generally, you can combine pivot_longer() with all the tidy selection functions to select columns, for example, you can use starts_with() or combine starts_with() and complement, !, or anything else the tidy selection language allows.

```
mean_income|>pivot_longer(
    names_to ="year",
    values_to ="income",
    starts_with("2")
)
```

```
## # A tibble: 25 × 3
##     country  year   income
##     <chr>    <chr>  <dbl>
##  1 Numenor  2001   123456
##  2 Numenor  2002   132654
##  3 Numenor  2003   321646
##  4 Numenor  2004   324156
##  5 Numenor  2005   325416
##  6 Westeros 2001   314256
##  7 Westeros 2002   432165
##  8 Westeros 2003   546123
##  9 Westeros 2004   465321
## 10 Westeros 2005   561423
## # . . . with 15 more rows
```

```
mean_income|>pivot_longer(
    names_to ="year",
    values_to ="income",
    !starts_with("c")
)
```

```
## # A tibble: 25 × 3
##     country  year   income
##     <chr>    <chr>  <dbl>
##  1 Numenor  2001   123456
```

```
##  2 Numenor   2002   132654
##  3 Numenor   2003   321646
##  4 Numenor   2004   324156
##  5 Numenor   2005   325416
##  6 Westeros  2001   314256
##  7 Westeros  2002   432165
##  8 Westeros  2003   546123
##  9 Westeros  2004   465321
## 10 Westeros  2005   561423
## # . . . with 15 more rows
```

The column names that go into the names_to column will be strings. If you need another type, such as logical or numeric values, you can achieve that by setting the names_transform argument. If we want all the years to be represented as integers, we could do this:

```
mean_income|>pivot_longer(
    names_to ="year",
    values_to ="income",
    !country,
    names_transform =as.integer
)
```

```
## # A tibble: 25 × 3
##    country  year   income
##    <chr>    <int>  <dbl>
##  1 Numenor  2001   123456
##  2 Numenor  2002   132654
##  3 Numenor  2003   321646
##  4 Numenor  2004   324156
##  5 Numenor  2005   325416
##  6 Westeros 2001   314256
##  7 Westeros 2002   432165
##  8 Westeros 2003   546123
##  9 Westeros 2004   465321
## 10 Westeros 2005   561423
## # . . . with 15 more rows
```

Because all the years are integers, the year column now contains integers. If we want to analyze this data set, we most likely want to interpret the years as integers.

If we wanted factors instead, though, we could also do that:

```
mean_income|>pivot_longer(
    names_to ="year",
    values_to ="income",
    !country,
    names_transform =as.factor
)
```

There is a similar `values_transform` argument if you need to transform the values that go into the values_to column.

If you work with functions that do not expect tidy data and rather expect to find related data in different columns, you can translate tidy data into that format using the `pivot_wider()` function. It is the reverse of `pivot_longer()`.[2]

```
tidy_income<-mean_income|>pivot_longer(
    names_to ="year",
    values_to ="income",
    !country
)

tidy_income|>pivot_wider(
    names_from ="year",
    values_from ="income"
)
```

```
## # A tibble: 5 × 6
##    country `2001` `2002` `2003` `2004` `2005`
##    <chr>   <dbl>  <dbl>  <dbl>  <dbl>  <dbl>
## 1 Numenor  123456 132654 321646 324156 325416
## 2 Westeros 314256 432165 546123 465321 561423
```

---

[2] The `pivot_longer()` and `pivot_wider()` functions transform tibbles in opposite directions in the sense that I have described here, but they are not truly each other's reverse. The types of values can change depending on the arguments provided to the functions, and one transformation does not always leave enough information that it can be reversed. Still, as a first approximation, it might be useful to think of them as each other's reverse.

```
## 3 Narnia    432156 342165 564123 543216 465321
## 4 Gondor    531426 321465 235461 463521 561423
## 5 Laputa     14235  34125  45123  51234  54321
```

The `names_from` and `values_from` arguments pick the values to use as column names and the values to put in those columns, respectively. They are the same arguments that you would use for `pivot_longer()`'s `names_to` and `values_to` arguments to translate the result back again.

Complex Column Encodings

Different variables are not always represented in separated columns. For example, it is not uncommon to have a column that contains a date, but that is really a day and a month and possibly a year. The best representation of a date is, of course, a date object, but for the sake of the example, let us say that we want to split a date into a day and a month column. You can do this using the `separate()` function.

```
tbl<-tribble(
    ~date,
    "11/5",
    "4/7",
    "21/12"
)
separate(tbl, date,into =c("day","month"))

## # A tibble: 3 × 2
##     day     month
##     <chr>   <chr>
## 1   11      5
## 2   4       7
## 3   21      12
```

The first argument is the data table, the second the column you want to split, and the `into` argument is a list of the columns the original date will be put into.

By default, the original column is removed, but if you want to keep it, you can use the `remove` argument.

```
separate(tbl, date,into =c("day","month"),remove =FALSE)
```

```
## # A tibble: 3 × 3
##    date  day   month
##    <chr> <chr> <chr>
## 1 11/5   11    5
## 2 4/7    4     7
## 3 21/12 21     12
```

After separating the fields in the data column, which is a column of strings, you get a day and a month column, and these also contain strings. To convert them into numbers, you can use the convert argument.

```
separate(tbl, date,into =c("day","month"),convert =TRUE)
```

```
## # A tibble: 3 × 2
##     day   month
##     <int> <int>
## 1   11        5
## 2   4         7
## 3   21       12
```

A column will be split by default on any non-alphanumeric character. This is why the / in the data is correctly used as the separator between the day and the month.[3] If your separator is something else, you can use the sep argument. The argument takes a regular expression, and the column will be split on this. For example, we can split on alphabet characters and spaces like this:

```
tbl<-tribble(
    ~date,
    "11th of month 5",
    "4th of month 7",
    "21st of month 12"
)
separate(tbl, date,into =c("day","month"),
        sep ="[[:alpha:][:space:]]+")
```

---

[3] Correctly, that is, if you use a sensible dd/mm notation and not something as silly as mm/dd.

```
## # A tibble: 3 × 2
##    day    month
##    <chr>  <chr>
## 1 11     5
## 2 4      7
## 3 21     12
```

The regular expression ([[:alpha:]]|[[:space:]])+ specifies any nonzero (specified by +) sequence of the character classes in (the outermost) square brackets, where we have put alphabet characters [:alpha:] and spaces [:space:].

The date format in this example is not something you would find in the wild, but if we have the date in this format

```
tbl<-tribble(
    ~date,
    "11th of May",
    "4th of July",
    "21st of December"
)
```

we cannot easily distinguish the separator string from the months. If the delimiter you need is not a single regular expression, you might be able to use the extract() function instead of separate(). With extract(), you can use regular expression groups to identify the fields to extract.

You need an expression with groups—those are subexpressions in parentheses—and what those groups match will be the value put in the new columns. For our dates, it could look like this:

```
tbl|>extract(
    col =date,
    into =c("day","month"),
    regex =paste0(
        "([[:digit:]]+)",         # First group, the day
        "[[:alpha:][:space:]]+",  # Stuff we ignore; will match "th of"
                                  and such
```

```
        "[[:space:]]",          # The final space before the month name
        "([[:alpha:]]+)"        # Second group, the month
    )
)

## # A tibble: 3 × 1
##   date
##   <chr>
## 1 11th of May
## 2 4th of July
## 3 21st of December
```

For both separate() and extract(), you can have more than two resulting columns. You can provide more than two names to the into parameter, and then the expression you use for separating and extracting should provide the right number of values, of course.

```
tbl|>extract(
    col =date,
    into =c("day","fluf","month"),
    regex =paste0(
        "([[:digit:]]+)",       # First group, the day
        "([[:alpha:]]+)",       # st, nd, rd, th, that kind of stuff
                                #   goes in fluf
        "[[:alpha:][:space:]]+",# stuff between day and month; not a group
                                #   so we throw it away
        "([[:alpha:]]+)"        # Third group, the month
    )
)

## # A tibble: 3 × 1
##   date
##   <chr>
## 1 11th of May
## 2 4th of July
## 3 21st of December
```

The reverse of separating is uniting. For example, let tbl2 contain a day and a month column. I'm building such a table in the following code:

```
tbl<-tribble(
    ~date,
    "11/5",
    "4/7",
    "21/12"
)
tbl2<-tbl|>separate(col =date,into =c("day","month"))
```

Having created such a tibble in tbl2, we can use the unite() function to undo separate() much as we found pivot_longer/wider() to be able to undo each other (with the caveats that things can get lost of course). The col argument should be the column we create and the following columns those we create it from.

```
tbl2|>unite(col ="date", day, month)
```

```
## # A tibble: 3 × 1
##    date
##    <chr>
## 1 11_5
## 2 4_7
## 3 21_12
```

By default, the separator in the new column will be _. This is not what we want for dates, but we can use the sep argument to change it.

```
tbl2|>unite(col ="date", day, month,sep ="/")
```

```
## # A tibble: 3 × 1
##    date
##    <chr>
## 1 11/5
## 2 4/7
## 3 21/12
```

If we want to keep the original columns around, we can set the remove argument to FALSE.

```
tbl2|>unite(
    col ="date", day, month,
    sep ="/",remove =FALSE
)
```

```
## # A tibble: 3 × 3
##   date    day    month
##   <chr>  <chr>  <chr>
## 1 11/5    11     5
## 2 4/7     4      7
## 3 21/12   21     12
```

Columns that contain more than one value do not always contain the same number of values. For example, we could have data such as the number of casualties per major group in WW1 and WW2.

```
military_casualties<-tribble(
    ~war,~groups,~deaths,

    'WW1',
    "Allied Powers/Central Powers",
    "5.7,4.0",

    'WW2',
    "Germany/Japan/USSR/British Empire/USA",
    "5.3,2.1,10.7,0.6,0.4"
)
```

The groupings were not the same in the two wars, so we cannot split the data into different columns. We can, however, divide it into more rows using separate_rows().

```
military_casualties|>separate_rows(
    groups, deaths,
    sep ="/|,"
)
```

```
## # A tibble: 7 × 3
##   war   groups         deaths
##   <chr> <chr>          <chr>
## 1 WW1   Allied Powers  5.7
## 2 WW1   Central Powers 4.0
## 3 WW2   Germany        5.3
## 4 WW2   Japan          2.1
## 5 WW2   USSR           10.7
## 6 WW2   British Empire 0.6
## 7 WW2   USA            0.4
```

The expression we use for sep says we should split on / or , (where the | in the string is regular expression speak for "or").

We might not want the deaths here to be strings, but with convert = TRUE we can turn them into numbers:

```
military_casualties|>separate_rows(
    groups, deaths,
    sep ="/|,",
    convert =TRUE
)
```

```
## # A tibble: 7 × 3
##   war   groups         deaths
##   <chr> <chr>          <dbl>
## 1 WW1   Allied Powers  5.7
## 2 WW1   Central Powers 4
## 3 WW2   Germany        5.3
## 4 WW2   Japan          2.1
## 5 WW2   USSR           10.7
## 6 WW2   British Empire 0.6
## 7 WW2   USA            0.4
```

Expanding, Crossing, and Completing

For some applications, it is useful to create all combinations of values from two or more columns—even those combinations that are missing from the data. For this, you can use the function expand():

```
tbl<-tribble(
   ~A,~B,~C,
    1,11,21,
    2,11,22,
    4,13,32
)
tbl|>expand(A, B)

## # A tibble: 6 × 2
##       A     B
##   <dbl> <dbl>
## 1     1    11
## 2     1    13
## 3     2    11
## 4     2    13
## 5     4    11
## 6     4    13
```

Here, we get all combinations of values from columns A and B while column C is ignored. If a column contains values from a larger set than what is found in the data, you can give expand() a vector of values:

```
tbl |> expand(A = 1:4, B)

## # A tibble: 8 × 2
##       A     B
##   <int> <dbl>
## 1     1    11
## 2     1    13
## 3     2    11
## 4     2    13
## 5     3    11
```

```
## 6      3      13
## 7      4      11
## 8      4      13
```

If you have vectors of values you want to combine this way, you can also use the crossing() function.

```
crossing(A = 1:3, B = 11:13)
```

```
## # A tibble: 9 × 2
##         A      B
##     <int>  <int>
## 1       1     11
## 2       1     12
## 3       1     13
## 4       2     11
## 5       2     12
## 6       2     13
## 7       3     11
## 8       3     12
## 9       3     13
```

If you only want the combinations found in the data, you can combine expand() with the nesting() function:

```
tbl <- tribble(
    ~A, ~B, ~C,
    1, 11, 21,
    2, 11, 22,
    2, 11, 12,
    4, 13, 42,
    4, 13, 32
)
expand(tbl, nesting(A, B))
```

```
## # A tibble: 3 × 2
##         A      B
##     <dbl>  <dbl>
```

```
## 1      1      11
## 2      2      11
## 3      4      13
```

This combination gives you all the combinations of the specified columns—all other columns are ignored—and only the unique combinations, unlike if you extracted the columns by indexing. This is useful if you have data with lots of duplications, but you only need to compute some statistics for the unique combinations.

If you need all combinations in a data frame, but accept missing values for those that are not present, you can use the complete() function. It makes sure you have all combinations and put missing values in the remaining columns for the combinations that are not in the original data.

```
complete(tbl, A = 1:4)
## # A tibble: 6 × 3
##        A      B      C
##    <dbl> <dbl> <dbl>
## 1      1     11     21
## 2      2     11     22
## 3      2     11     12
## 4      3     NA     NA
## 5      4     13     42
## 6      4     13     32

complete(tbl, B = 11:13)

## # A tibble: 6 × 3
##        B      A      C
##    <dbl> <dbl> <dbl>
## 1     11      1     21
## 2     11      2     22
## 3     11      2     12
## 4     12     NA     NA
## 5     13      4     42
## 6     13      4     32

complete(tbl, A = 1:4, B = 11:13)
```

```
## # A tibble: 14 × 3
##        A     B     C
##    <dbl> <dbl> <dbl>
## 1      1    11    21
## 2      1    12    NA
## 3      1    13    NA
## 4      2    11    22
## 5      2    11    12
## 6      2    12    NA
## 7      2    13    NA
## 8      3    11    NA
## 9      3    12    NA
## 10     3    13    NA
## 11     4    11    NA
## 12     4    12    NA
## 13     4    13    42
## 14     4    13    32
```

Missing Values

When data has missing values, it often requires application domain knowledge to deal with it correctly. The `tidyr` package has some rudimentary support for cleaning data with missing values; you might need more features to deal with missing values properly, and you are likely to find those in the `dplyr` package (see Chapter 8).

The simplest way to handle missing values is to get rid of it. A crude approach is to remove all observations with one or more missing variable values. The `drop_na()` function does exactly that.

```
mean_income <- tribble(
    ~country, ~`2002`, ~`2003`, ~`2004`, ~`2005`,
    "Numenor", 123456, 132654, NA, 324156,
    "Westeros", 314256, NA, NA, 465321,
    "Narnia", 432156, NA, NA, NA,
    "Gondor", 531426, 321465, 235461, 463521,
    "Laputa", 14235, 34125, 45123, 51234,
)
```

```
drop_na(mean_income)
```

```
## # A tibble: 2 × 5
##    country `2002` `2003` `2004` `2005`
##    <chr>    <dbl>  <dbl>  <dbl>  <dbl>
## 1 Gondor  531426 321465 235461 463521
## 2 Laputa   14235  34125  45123  51234
```

In this example, we lose a country entirely if we have missing data for a single year. This is unlikely to be acceptable and is another good argument for tidy data. If we reformat the table and drop missing values, we will only remove the observations where the income is missing for a given year.

```
mean_income |> pivot_longer(
    names_to = "year",
    values_to = "mean_income",
    -country
) |> drop_na()
```

```
## # A tibble: 14 × 3
##    country  year  mean_income
##    <chr>    <chr>        <dbl>
## 1  Numenor  2002        123456
## 2  Numenor  2003        132654
## 3  Numenor  2005        324156
## 4  Westeros 2002        314256
## 5  Westeros 2005        465321
## 6  Narnia   2002        432156
## 7  Gondor   2002        531426
## 8  Gondor   2003        321465
## 9  Gondor   2004        235461
## 10 Gondor   2005        463521
## 11 Laputa   2002         14235
## 12 Laputa   2003         34125
## 13 Laputa   2004         45123
## 14 Laputa   2005         51234
```

Sometimes, you can replace missing data with some appropriate data. For example, we could replace missing data in the `mean_income` table by setting each value to the mean of its column. The `replace_na()` function does this. It takes a data frame as its first argument and a list specifying replacements as its second argument. This list should have a name for each column you want to replace missing values in, and the value assigned to these names should be the values with which you wish to replace.

```
replace_na(mean_income, list(
    `2003` = mean(mean_income$`2003`, na.rm = TRUE)
))
```

```
## # A tibble: 5 × 5
##   country   `2002` `2003`  `2004`  `2005`
##   <chr>      <dbl>  <dbl>   <dbl>   <dbl>
## 1 Numenor   123456 132654      NA  324156
## 2 Westeros  314256 162748      NA  465321
## 3 Narnia    432156 162748      NA      NA
## 4 Gondor    531426 321465  235461  463521
## 5 Laputa     14235  34125   45123   51234
```

```
replace_na(mean_income, list(
    `2003` = mean(mean_income$`2003`, na.rm = TRUE),
    `2004` = mean(mean_income$`2004`, na.rm = TRUE),
    `2005` = mean(mean_income$`2005`, na.rm = TRUE)
))
```

```
## # A tibble: 5 × 5
##   country   `2002` `2003` `2004`  `2005`
##   <chr>      <dbl>  <dbl>  <dbl>   <dbl>
## 1 Numenor   123456 132654 140292  324156
## 2 Westeros  314256 162748 140292  465321
## 3 Narnia    432156 162748 140292  326058
## 4 Gondor    531426 321465 235461  463521
## 5 Laputa     14235  34125  45123   51234
```

In this example, using the mean per year is unlikely to be useful; it would make more sense to replace missing data with the mean for each country. This is not immediately possible with tidyr functions, though. With `dplyr`, we have the tools for it, and I return to it in Chapter 8.

A final function for managing missing data is the `fill()` function. It replaces missing values with the value above them in their column (or below them if you set the argument `.direction` to "up". Imagine that we have a table of income per quarter of each year, but the year is only mentioned for the first quarter.

```
tbl <- read_csv(
    "year, quarter, income
    2011, Q1, 13
    , Q2, 12
    , Q3, 14
    , Q4, 11
    2012, Q1, 12
    , Q2, 14
    , Q3, 15
    , Q4, 17",
show_col_types = FALSE
)
spec(tbl)

## cols(
##   year = col_double(),
##   quarter = col_character(),
##   income = col_double()
## )
```

We can repeat the years downward through the quarters with `fill()`:

```
fill(tbl, year)

## # A tibble: 8 × 3
##     year quarter income
##    <dbl> <chr>    <dbl>
## 1   2011 Q1          13
```

```
## 2   2011 Q2         12
## 3   2011 Q3         14
## 4   2011 Q4         11
## 5   2012 Q1         12
## 6   2012 Q2         14
## 7   2012 Q3         15
## 8   2012 Q4         17
```

Nesting Data

A final functionality that tidyr provides is nesting data. A tibble entry usually contains simple data, such as numbers or strings, but it can also hold complex data, such as other tables.

Consider this table:

```
tbl<-tribble(
    ~A,~B,~C,
    1,11,21,
    1,11,12,
    2,42,22,
    2,15,22,
    2,15,32,
    4,13,32
)
```

From this table, we can create one that, for each value in the A column, contains a table of B and C for that A value.

```
nested_tbl <- tbl |> nest(BC = c(B, C))
nested_tbl

## # A tibble: 3 × 2
##       A BC
##   <dbl> <list>
## 1     1 <tibble [2 × 2]>
## 2     2 <tibble [3 × 2]>
## 3     4 <tibble [1 × 2]>
```

```
nested_tbl[[1,2]]
```

```
## [[1]]
## # A tibble: 2 × 2
##       B     C
##   <dbl> <dbl>
## 1    11    21
## 2    11    12
```

The column that contains the nested table is named BC because we used a named argument with that name, BC = c(B, C).

The reverse operation is unnest():

```
nested_tbl |> unnest(cols = c(BC))
```

```
## # A tibble: 6 × 3
##       A     B     C
##   <dbl> <dbl> <dbl>
## 1     1    11    21
## 2     1    11    12
## 3     2    42    22
## 4     2    15    22
## 5     2    15    32
## 6     4    13    32
```

Nesting is not a typical operation in data analysis, but it is there if you need it, and in the packages in Chapter 12, we see some usages.

## CHAPTER 6

Pipelines: `magrittr`

Data analysis consists of several steps, where your data moves through different stages of transformations and cleaning before you finally get to model construction. In practical terms, this means that your R code will consist of a series of function calls where the output of one is the input of the next. The pattern is typical, but a straightforward implementation of it has several drawbacks. The Tidyverse has for many years provided a "pipe operator" to alleviate this, and with R 4.1, there is also a built-in operator in the language itself. Although this chapter is mainly about the `magrittr` pipe operator, %>%, I will describe the native |> operator as well and point out a few places where they differ in their behavior.

The native operator, |>, is readily available in any R version greater than 4.1.0. You can get the pipe operator implemented in the `magrittr` package if you load the `tidyverse` package:

```
library(tidyverse)
```

or explicitly load the package:

```
library(magrittr)
```

If you load `magrittr` through the `tidyverse` package, you will get the most common pipe operator, %>%, but not alternative ones—see the following section. For those, you need to load `magrittr`.

The Problem with Pipelines

Consider this data from Chapter 5.

```
write_csv(
    tribble(
        ~country,   ~`2002`, ~`2003`, ~`2004`, ~`2005`,
```

© Thomas Mailund 2022

T. Mailund, *R 4 Data Science Quick Reference*, https://doi.org/10.1007/978-1-4842-8780-4_6

```
    "Numenor",  123456,  132654,        NA,  324156,
    "Westeros", 314256,  NA,            NA,  465321,
    "Narnia",   432156,  NA,            NA,       NA,
    "Gondor",   531426,  321465,  235461,  463521,
    "Laputa",    14235,   34125,   45123,   51234,
  ),
  "data/income.csv"
)
```

I have written it to a file, data/income.csv, to make the following examples easier
to read.

In Chapter 5, we reformatted and cleaned this data using pivot_longer() and
drop_na():

```
mydata <- read_csv("data/income.csv",
                   col_types = "cdddd")
mydata <- pivot_longer(
    data = mydata,
    names_to = "year",
    values_to = "mean_income",
    !country
)
mydata <- drop_na(mydata)
```

This is a typical pipeline, although a short one. It consists of three steps: (1) read the
data into R, (2) tidy the data, and (3) clean the data.

There is nothing inherently wrong with code like this. Each step is easy to read; it is
a function call that transforms your data from one form to another. The pipeline nature
of the code, however, is not explicit. We can read the code and see that the output of one
function is the input of the next, but what if it isn't?

Consider this:

```
mydata <- read_csv("data/income.csv",
                   col_types = "cdddd")
data <- pivot_longer(...) # Are we assigning to the right name?
mydata <-drop_na(mydata) # Input from read_csv() not pivot_longer()
```

The result of `pivot_longer()` is assigned to the variable `data`, and then `drop_na()` is called on `mydata`. If our intent were the preceding pipeline, this would be a mistake. But maybe we wanted the result of `pivot_longer()` to be a separate data table needed for some downstream analysis.

A series of transformation steps that constitute a pipeline can only be recognized as such by following programming conventions, and unless the code is well documented and well understood, we cannot immediately see from the code whether assigning to `data` instead of `mydata` is an error.

We can make pipelines more explicit. If the output of one function should be the input of the next, we can nest the function calls:

```
mydata <- drop_na(
    pivot_longer(
        data = read_csv("data/income.csv",
                        col_types = "cdddd"),
        names_to = "year",
        values_to = "mean_income",
        !country
    )
)
```

This makes the pipeline intention explicit, but the code very hard to read; you have to work out the pipeline from the innermost function call to the outermost.

The `magrittr` package gives us notation for explicitly describing specifying readable pipelines.

Pipeline Notation

The pipeline operators, |> and %>%, introduce syntactic sugar for function calls.[1]
The code

```
x %>% f()
```

or

```
x |> f()
```

is equivalent to

```
f(x)
```

and

```
x %>% f() %>% g() %>% h()
```

or

```
x |> f() |> g() |> h()
```

is equivalent to

```
h(g(f(x)))
```

So, by the way, is

```
x %>% f %>% g %>% h
```

---

[1] The magrittr pipeline operator is "syntactic sugar" in the sense that it introduces a more readable notation for a function call. In compiled languages, syntactic sugar would be directly translated into the original form and have exactly the same runtime behavior as the sugar-free version. The built-in pipeline operator, |>, available after R 4.1, works this way. When R parses your source code, it will translate an expression such as x |> f() into f(x), and there is practically no difference between using the two syntaxes for calling a function. With magrittr's operator, however, %>% is just a function as any other R function and calling it incurs a runtime cost. The %>% operator is a complex function, so compared to explicit function calls, it is slow to use it. This is rarely an issue, though, since the expensive data processing is done inside the function and not in function calls. The extra power that magrittr's operator provides, compared to the new built-in operator, often makes it a better choice. There is nothing wrong with mixing the two operators, however, and use |> for simple (and faster) function calls and %>% for more elaborate expressions. Keep in mind, though, that they do differ in some use cases, so you cannot always use them as drop-in replacements.

but not

```
x |> f |> g |> h
```

The native operator needs to have a function *call*, f(), on the right-hand side of the operator, while the magrittr will accept either a function *call*, f(), or just a function *name*, x %>% f. You can also give %>% an anonymous function:

```
"hello, "%>% (\(y){ paste(y, "world!")})
```

```
## [1] "hello, world!"
```

but |> still needs a function *call*, so if you give it an anonymous function, you must call the function as well (add () after the function definition).

```
"hello," |> (\(y){ paste(y, "world!")})()
```

```
## [1] "hello, world!"
```

In either case, you *must* put the function definition in parentheses. This will be an error:

```
x |> \(y) { ... }
```

This is because both operators are syntactic sugar that translates x %>% f(), x %>% f, or x |> f() into f(x) and the function definition, function(y) { ... } or \(y) { ... }, *is* a function call. The left-hand side of the pipeline is put into the function definition:

```
x |> \(y) { ... }
```

is translated into

```
\(x, f) { ... }
```

for both operators. Putting the function definition in parentheses prevents this; while function(y) { ... } is a function call (to the function called function), (function(y) { ... }) isn't. That is, when the function definition is in parentheses, it is no longer syntactically a function call (but an expression in parentheses). If it is not a call, |> will not accept it as the right-hand side, but %>% will. For %>%, it still isn't a call, just something that evaluates to one, so the function definition isn't modified, but the result gets the left-hand side as an argument.

Anyway, that got technical, but it is something you need to know if you use anonymous functions in pipelines.

With the %>% operator, you do not need the parentheses for the function calls, but most prefer them to make it clear that we are dealing with functions, and the syntax is then still valid if you change the code to the |> operator. Also, if your functions take more than one argument, parentheses are needed, so if you always include them, it gives you a consistent notation.

The nested function call pipeline from earlier

```
mydata <- drop_na(pivot_longer(read_csv(. . . )))
```

can thus be rewritten in pipe form

```
mydata <- read_csv(. . . ) %>% pivot_longer(. . . ) %>% drop_na()
```

or, with all the arguments included, as

```
mydata <- read_csv("data/income.csv", col_types = "cdddd") %>%
    pivot_longer(names_to = "year", values_to = "mean_income", !country) %>%
    drop_na()
```

The pipeline notation combines the readability of single function call steps with explicit notation for data processing pipelines.

Pipelines and Function Arguments

The pipe operator is left-associative, that is, it evaluates from left to right, and the expression lhs %>% rhs expects the left-hand side (lhs) to be data and the right-hand side (rhs) to be a function or a function call (and always a function call for |>). The result will be rhs(lhs). The output of this function call is the left-hand side of the next pipe operator in the pipeline or the result of the entire pipeline. If the right-hand side function takes more than one argument, you provide these in the rhs expression. The expression

```
lhs %>% rhs(x, y, z)
```

will be evaluated as the function call

```
rhs(lhs, x, y, z)
```

By default, the left-hand side is given as the first argument to the right-hand side function. The arguments that you explicitly write on the right-hand side expression are the additional function parameters.

With the `magrittr` operator, %>%, but not the |> operator, you can change the default input position using the special variable . (dot). If the right-hand side expression has . as a parameter, then that is where the left-hand side value goes. All the following four pipelines are equivalent:

```
mydata <- read_csv("data/income.csv",
                   col_types = "cdddd") |>
    pivot_longer(names_to = "year", values_to = "mean_income", !country) |>
    drop_na()

mydata <- read_csv("data/income.csv",
                   col_types = "cdddd") %>%
    pivot_longer(names_to = "year", values_to = "mean_income", !country) %>%
    drop_na()

mydata <- read_csv("data/income.csv",
                   col_types = "cdddd") %>%
    pivot_longer(data = ., names_to = "year", values_to = "mean_income",
    !country) %>%
    drop_na()

mydata <- read_csv("data/income.csv",
                   col_types = "cdddd") %>%
    pivot_longer(names_to = "year", values_to = "mean_income", !country,
    data = .) %>%
    drop_na()
```

You can use a dot more than once on the right-hand side:

```
rnorm(5) %>% tibble(x = ., y = .)

## # A tibble: 5 × 2
##        x        y
##     <dbl>    <dbl>
## 1 -0.951  -0.951
## 2  0.0206  0.0206
## 3  1.14    1.14
## 4 -0.172  -0.172
## 5  0.0795  0.0795
```

You can also use it in nested function calls.

```
rnorm(5) %>% tibble(x = ., y = abs(.))
```

```
## # A tibble: 5 × 2
##            x        y
##        <dbl>    <dbl>
## 1   1.61       1.61
## 2   1.86       1.86
## 3  -0.211     0.211
## 4  -0.00352  0.00352
## 5  -0.522     0.522
```

If the dot is *only* found in nested function calls, however, magrittr will still add it as the first argument to the right-hand side function.

```
rnorm(5) %>% tibble(x = sin(.), y = abs(.))
```

```
## # A tibble: 5 × 3
##           .       x      y
##       <dbl>   <dbl>  <dbl>
## 1   1.10     0.893   1.10
## 2  -0.155   -0.155   0.155
## 3   0.102    0.102   0.102
## 4   0.627    0.587   0.627
## 5   0.280    0.276   0.280
```

To avoid this, you can put the right-hand side expression in curly brackets:

```
rnorm(5) %>% { tibble(x = sin(.), y = abs(.)) }
```

```
## # A tibble: 5 × 2
##            x       y
##        <dbl>   <dbl>
## 1   0.661    0.723
## 2   0.0588   0.0588
## 3   0.960    1.86
## 4   0.820    0.961
## 5  -0.366    0.375
```

In general, you can put expressions in curly brackets as the right-hand side of a pipe operator and have `magrittr` evaluate them. Think of it as a way to write one-parameter anonymous functions. The input variable is the dot, and the expression inside the curly brackets is the body of the function.

Function Composition

It is trivial to write your own functions such that they work well with pipelines. All functions map input to output (ideally with no side effects) so all functions can be used in a pipeline. If the key input is the first argument of a function you write, the default placement of the left-hand side value will work—so that is preferable—but otherwise you can explicitly use the dot.

When writing pipelines, however, often you do not need to write a function from scratch. If a function is merely a composition of other function calls—it is a pipeline function itself—we can define it as such.

In mathematics, the function composition operator, $\circ$, defines the composition of functions f and g, $g \circ f$, to be the function

$$(g \circ f)(x) = g(f(x))$$

Many functional programming languages encourage you to write functions by combining other functions and have operators for that. It is not frequently done in R, and while you can implement function composition, there is no built-in operator. The `magrittr` package, however, gives you this syntax:

```
h <- . %>% f() %>% g()
```

This defines the function `h`, such that `h(x) = g(f(x))`.

If we take the tidy-then-clean pipeline we saw earlier, and imagine that we need to do the same for several input files, we can define a pipeline function for this as

```
pipeline <- . %>%
    pivot_longer(names_to = "year", values_to = "mean_income", !country) %>%
    drop_na()
pipeline

## Functional sequence with the following components:
##
##  1. pivot_longer(., names_to = "year", values_to = "mean_income", !country)
```

```
##  2. drop_na(.)
##
## Use 'functions' to extract the individual functions.
```

Other Pipe Operations

There are three other pipe operators in magrittr. These are not imported when you import the tidyverse package, so to get access to them, you have to import magrittr explicitly.

```
library(magrittr)
```

The %<>% operator is used to run a pipeline starting with a variable and ending by assigning the result of the pipeline back to that variable.

The code

```
mydata <- read_csv("data/income.csv", col_types = "cdddd")
mydata %<>% pipeline()
```

is equivalent to

```
mydata <- read_csv("data/income.csv", col_types ="cdddd")
mydata <- mydata %>% pipeline()
```

This reassignment operator behaves similarly to the stepwise pipeline convention we considered at the start of the chapter, but it makes explicit that we are updating an existing variable. You cannot accidentally assign the result to a different variable.

If your right-hand side is an expression in curly brackets, you can refer to the input through the dot variable:

```
mydata <- tibble(x = rnorm(5), y = rnorm(5))
mydata %>% { .$x - .$y }
```

```
## [1] 0.3597903 1.8108120 2.3444173 3.1367824
## [5] 0.2507438
```

If, as here, the input is a data frame and you want to access its columns, you need the notation .$ to access them. The %$% pipe operator opens the data frame for you so you can refer to the columns by name.

```
mydata %$% { x - y }
```

```
## [1] 0.3597903 1.8108120 2.3444173 3.1367824
## [5] 0.2507438
```

The *tee* pipe operator, %T>%, behaves like the regular pipe operator, %>%, except that it returns its input rather than the result of calling the function on its right-hand side. The regular pipe operation x %>% f() will return f(x), but the tee pipe operation x %T>% f() will call f(x) but return x. This is useful for calling functions with side effects as a step inside a pipeline, such as saving intermediate results to files or plotting data.

If you call a function in a usual pipeline, you will pass the result of the function call on to the next function in the pipeline. If you, for example, want to plot intermediate data, you might not be so lucky that the plotting function will return the data. The ggplot2 functions will not (see Chapter 13).

If you want to plot intermediate values, you need to save the data in a variable, save it, and then start a new pipeline with the data as input.

```
tidy_income <- read_csv("data/income.csv", col_types = "cdddd") |>
    pivot_longer(names_to = "year", values_to = "mean_income", !country)

# Summarize and visualize data before we continue
summary(tidy_income)
ggplot(tidy_income, aes(x = year, y =mean_income)) + geom_point()

# Continue processing
tidy_income |>
    drop_na() |>
    write_csv("data/tidy-income.csv")
```

With the tee operator, you can call the plotting function and continue the pipeline.

```
mydata <- read_csv("data/income.csv", col_types = "cdddd") %>%
    pivot_longer(names_to = "year", values_to = "mean_income", !country) %T>%
    # Summarize and then continue
    { print(summary(.)) } %T>%
    # Plot and then continue
    { print(ggplot(., aes(x = year, y = mean_income)) + geom_point()) }%>%
    drop_na() %>% write_csv("data/tidy-income.csv")
```

## CHAPTER 7

Functional Programming: `purrr`

A pipeline-based approach to data processing necessitates a functional programming approach. After all, pipelines are compositions of functions, and loops and variable assignment do not work well with pipelines. You are not precluded from imperative programming, but you need to wrap it in functions.

The package `purrr` makes functional programming easier. As with the other packages in this book, it will be loaded if you import `tidyverse`, but you can load it explicitly with

```
library(purrr)
```

I will not describe all the functions in the `purrr` package, there are many and more could be added between the time I write this and the time you read it, but I will describe the functions you will likely use often.

Many of the examples in this chapter are so trivial that you would not use `purrr` for them. Many reduce to vector expressions, and vector expressions are both more straightforward to read and faster to evaluate. However, applications where you cannot use vector expressions are more involved, and I do not want the examples to overshadow the syntax.

© Thomas Mailund 2022
T. Mailund, *R 4 Data Science Quick Reference*, https://doi.org/10.1007/978-1-4842-8780-4_7

General Features of purrr Functions

The functions in the purrr package are "high-order" functions, which means that the functions either take other functions as input or return functions when you call them (or both). Most of the functions take data as their first argument and return modified data. Thus, they are immediately usable with pipelines. As additional arguments, they will accept one or more functions that specify how you want to translate the input into the output.

Other functions do not transform data but instead modify functions. These are used, so you do not need to write functions for the data transformation functions explicitly; you can use a small set of functions and adapt them where needed.

Filtering

One of the most straightforward functional programming patterns is filtering. Here, you use a high-order filter function. This function takes a predicate function, that is, a function that returns a logical value, and then returns all elements where the predicate evaluates to TRUE (the function keep()) or all elements where the predicate evaluates to FALSE (the function discard()).

```
is_even <- function(x) x %% 2 == 0
1:6 |> keep(is_even)

## [1] 2 4 6

1:6 |> discard(is_even)

## [1] 1 3 5
```

When you work with predicates, the negate() function can come in handy. It changes the truth value of your predicate, such that if your predicate, p, returns TRUE, then negate(p) returns FALSE and vice versa.

```
1:6 |> keep(negate(is_even))

## [1] 1 3 5

1:6 |> discard(negate(is_even))

## [1] 2 4 6
```

Since you already have complementary functions in keep() and discard(), you would not use negate() for filtering, though.

A special-case function, compact(), removes NULL elements from a list.

```
y <- list(NULL, 1:3, NULL)
y |> compact()
```

```
## [[1]]
## [1] 1 2 3
```

If you access attributes on objects, and those are not set, R will give you NULL as a result, and for such cases, compact() can be useful.

If we define lists

```
x <- y <- 1:3
names(y) <- c("one", "two", "three")
```

then y's values will be named, while x's will not. If we try to get the names from x, we will get NULL.

```
names(x)
```

```
## NULL
```

```
names(y)
```

```
## [1] "one"    "two"    "three"
```

Using compact() with names(), we will discard x.

```
z <- list(x = x, y = y)
z |> compact(names)
```

```
## $y
##   one   two   three
```

```
##      1      2       3
```

Mapping

Mapping a function, f, over sequences $x = x_1, x_2, ..., x_n$ returns a new sequence of the same length as the input but where each element is an application of the function: $f(x_1)$, $f(x_2), ..., f(x_n)$.

The function `map()` does this and returns a `list` as output. Lists are the generic sequence data structure in R since they can hold all types of R objects.

```
is_even <- function(x) x %% 2 == 0
1:4 |> map(is_even)
```

```
## [[1]]
## [1] FALSE
##
## [[2]]
## [1] TRUE
##
## [[3]]
## [1] FALSE
##
## [[4]]
## [1] TRUE
```

Often, we want to work with vectors of specific types. For all the atomic types, `purrr` has a specific mapping function. The function for logical values is named `map_lgl()`.

```
1:4 |> map_lgl(is_even)
```

```
## [1] FALSE TRUE FALSE TRUE
```

With something as simple as this example, you should not use `purrr`. Vector expressions are faster and easier to use.

```
1:4 %% 2 == 0
```

```
## [1] FALSE TRUE FALSE TRUE
```

You cannot always use vector expressions, however. Say you want to sample *n* elements from a normal distribution, for a sequence of different *n* values, and then calculate the standard error of the mean. The vector expression

```
sd(rnorm(n = n)) / sqrt(n) # not SEM!
```

does not compute this. If you give rnorm() a sequence for its n parameter, it takes the *length* of the input as the number of elements to sample.

```
n <- seq.int(100, 1000, 300) # the different n value we want
n
```

```
## [1]  100  400  700 1000
```

```
# Are we sampling for each of our n values?
rnorm(n) # no, but length(n) is used for the number of samples in rnorm()
```

```
## [1]  0.6691846 -0.4554893 -0.8175801  1.7794664
```

When you have a function that is not vectorized, you can use a map function to apply it on all elements in a list.

```
sem <- function(n) sd(rnorm(n = n)) / sqrt(n)
n |> map_dbl(sem)
```

```
## [1] 0.10197308 0.05138140 0.03673840 0.03057261
```

Here, I used map_dbl() to get doubles (numerics) as output. The functions map_chr() and map_int() will give you strings and integers instead.

```
1:3 %>% map_dbl(identity) %T>% print() %>% class()
```

```
## [1] 1 2 3
```

```
## [1] "numeric"
```

```
1:3 %>% map_chr(identity) %T>% print() %>% class()
```

```
## [1] "1" "2" "3"
```

```
## [1] "character"
```

```
1:3 %>% map_int(identity) %T>% print() %>% class()
```

```
## [1] 1 2 3
```

```
## [1] "integer"
```

The different map functions will give you an error if the function you apply does not return values of the type the function should. Sometimes, a map function will do type conversion, as before, but not always. It will usually be happy to convert in a direction that doesn't lose information, for example, from logical to integer and integer to double, but not in the other direction.

For example, map_lgl() will not convert from integers to logical if you give it the identity() function, even though you can convert these types. Similarly, you can convert strings to integers using as.numeric(), but map_dbl() will not do so if we give it identity(). The mapping functions are more strict than the base R conversion functions. So you should use functions that give you the right type.

The map_dfr() and map_dfc() functions return data frames (tibbles). The function you map should return data frames, and these will be combined, row-wise with map_dfr() and column-wise with map_dfc().

```
x <- tibble(a = 1:2, b = 3:4)
list(a = x, b = x) |> map_dfr(identity)
```

```
## # A tibble: 4 × 2
##        a     b
##    <int> <int>
## 1      1     3
## 2      2     4
## 3      1     3
## 4      2     4
```

```
list(a = x, b = x) |> map_dfc(identity)
```

```
## New names:
## • `a` -> `a...1`
## • `b` -> `b...2`
## • `a` -> `a...3`
## • `b` -> `b...4`
```

```
## # A tibble: 2 × 4
##   a...1 b...2 a...3 b...4
##   <int> <int> <int> <int>
## 1     1     3     1     3
## 2     2     4     2     4
```

The `map_df()` function does the same as `map_dfr()`:

```
list(a = x, b = x) |> map_df(identity)
```

```
## # A tibble: 4 × 2
##       a     b
##   <int> <int>
## 1     1     3
## 2     2     4
## 3     1     3
## 4     2     4
```

You do not need to give the data frame functions data frames as input, as long as the function you apply to the input returns data frames. This goes for all the map functions. They will accept any sequence input; they only restrict and convert the output.

If the items you map over are sequences themselves, you can extract elements by index; you do not need to provide a function to the map function.

```
x <- list(1:3, 4:6)
x |> map_dbl(1)
```

```
## [1] 1 4
```

```
x |> map_dbl(3)
```

```
## [1] 3 6
```

If the items have names, you can also extract values using these.

```
x <- list(
    c(a = 42, b = 13),
    c(a = 24, b = 31)
)
x |> map_dbl("a")
```

```
## [1] 42 24
```

```
x |> map_dbl("b")
```

```
## [1] 13 31
```

This is mostly used when you map over data frames.

```
a <- tibble(foo = 1:3, bar = 11:13)
b <- tibble(foo = 4:6, bar = 14:16)
ab <- list(a = a, b = b)
ab |> map("foo")
```

```
## $a
## [1] 1 2 3
##
## $b
## [1] 4 5 6
```

Related to extracting elements by name or index, you can apply functions to different depths of the input using map_depth(). Depth zero is the list itself, so mapping over this depth is the same as applying the function directly on the input.

```
ab |> map_depth(0, length)
```

```
## [1] 2
```

```
ab |> length()
```

```
## [1] 2
```

Depth 1 gives us each element in the sequence, so this behaves like a normal map. Depth 2 provides us with a map over the nested elements. Consider the list ab earlier. The top level, depth 0, is the list. Depth 1 is the data frames a and b. Depth 2 is the columns in these data frames. Depth 3 is the individual items in these columns.

```
ab |> map_depth(1, sum) |> unlist()
```

```
##  a  b
## 42 60
```

```
ab |> map_depth(2, sum) |> unlist()
```

```
## a.foo a.bar b.foo b.bar
##     6    36    15    45
```

```
ab |> map_depth(3, sum) |> unlist()
```

```
## a.foo1 a.foo2 a.foo3 a.bar1 a.bar2 a.bar3 b.foo1
##      1      2      3     11     12     13      4
## b.foo2 b.foo3 b.bar1 b.bar2 b.bar3
##      5      6     14     15     16
```

If you only want to apply a function to some of the elements, you can use map_if().
It takes a predicate and a function and applies the function to those elements where the
predicate is true. It returns a list, but you can convert it if you want another type.

```
is_even <- function(x) x %% 2 == 0
add_one <- function(x) x + 1
1:6 |> map_if(is_even, add_one) |> as.numeric()
```

```
## [1] 1 3 3 5 5 7
```

Notice that this is different from combining filtering and mapping; that combination
would remove the elements that do not satisfy the predicate.

```
1:6 |> keep(is_even) |> map_dbl(add_one)
```

```
## [1] 3 5 7
```

With map_if(), you keep all elements, but the function is only applied to some
of them.

If you want to apply one function to the elements where the predicate is true and
another to the elements where it is false, you can prove a function to the .else element:

```
add_two <- function(x) x + 2
1:6 |>
    map_if(is_even, add_one, .else = add_two) |>
    as.numeric()
```

```
## [1] 3 3 5 5 7 7
```

If you know which indices you want to apply the function to, instead of a predicate they must satisfy, you can use map_at(). This function takes a sequence of indices instead of the predicate but otherwise works the same as map_if().

```
1:6 |> map_at(2:5, add_one) |> as.numeric()
```

```
## [1] 1 3 4 5 6 6
```

If you map over a list, x, then your function will be called with the elements in the list, x[[i]]. If you want to get the elements wrapped in a length-one list, that is, use indexing x[i], you can use lmap().

```
list(a = 1:3, b = 4:6) |> map(print) |> invisible()
```

```
## [1] 1 2 3
## [1] 4 5 6
```

```
list(a = 1:3, b = 4:6) |> lmap(print) |> invisible()
```

```
## $a
## [1] 1 2 3
##
## $b
## [1] 4 5 6
```

The function you apply must always return a list, and lmap() will concatenate them.

```
f <- function(x) list("foo")
1:2 |> lmap(f)
```

```
## [[1]]
## [1] "foo"
##
## [[2]]
## [1] "foo"
```

```
f <- function(x) list("foo", "bar")
1:2 |> lmap(f)
```

```
## [[1]]
## [1] "foo"
```

```
##
## [[2]]
## [1] "bar"
##
## [[3]]
## [1] "foo"
##
## [[4]]
## [1] "bar"
```

For example, while you can get the length of elements in a list using map() and length()

```
list(a = 1:3, b = 4:8) |> map(length) |> unlist()
```

```
## a b
## 3 5
```

you will get an error if you try the same using lmap(). This is because length() returns a numeric and not a list. You need to wrap length() with list() so the result is the length in a (length one) list.

```
wrapped_length <- function(x) {
    x |> length() |>          # get the length of x (result will be numeric)
        list() |>             # go back to a list
        set_names(names(x))   # and give it the original name
}
list(a = 1:3, b = 4:8) |> lmap(wrapped_length) |> unlist()
```

```
## a b
## 1 1
```

If it surprises you that the lengths are one here, remember that the function is called with the length-one lists at each index. If you want the length of what they contain, you need to extract that.

```
wrapped_length <- function(x) {
    x |> pluck(1) |>          # pluck(x, 1) from purrr does that same
                              as x[[1]]
```

```
    length() |>          # now we have the underlying vector and get
                         the length
    list() |>            # go back to a list
    set_names(names(x))  # and give it the original name
}
list(a = 1:3, b = 4:8) %>% lmap(wrapped_length) %>% unlist()
```

```
## a b
## 3 5
```

If you want to extract the nested data, though, you probably want map() and not lmap().

The functions lmap_if() and lmap_at() work as map_if() and map_at() except for how they index the input and handle the output as lists to be concatenated.

Sometimes, we only want to call a function for its side effect. In that case, you can pipe the result of a map into invisible(). The function walk() does that for you, and using it makes it explicit that this is what you want, but it is simply syntactic sugar for map() + invisible().

```
1:3 |> map(print) |> invisible()
```

```
## [1] 1
## [1] 2
## [1] 3
```

```
1:3 |> walk(print)
```

```
## [1] 1
## [1] 2
## [1] 3
```

If you need to map over multiple sequences, you have two choices of map functions to choose from. Some functions map over exactly two sequences. For each of the map() functions, there are similar map2() functions. These take two sequences as the first two arguments.

```
x <- 1:3
y <- 3:1
map2_dbl(x, y, `+`)
```

```
## [1] 4 4 4
```

You can also create lists of sequences and use the pmap() functions.

```
list(x, y) |> pmap_dbl(`+`)
```

```
## [1] 4 4 4
```

There are the same type-specific versions as there are for map() and map2(), but with the pmap() functions, you can map over more than one or two input sequences.

```
z <- 4:6
f <- function(x, y, z) x + y - z
list(x, y, z) |> pmap_dbl(f)
```

```
## [1] 0 -1 -2
```

If you need to know the indices for each value you map over, you can use the imap() variations. When you use these to map over a sequence, your function needs to take two arguments where the first argument is the sequence value and the second the value's index in the input.

```
x <- c("foo", "bar", "baz")
f <- function(x, i) paste0(i, ": ", x)
x |> imap_chr(f)
```

```
## [1] "1: foo" "2: bar" "3: baz"
```

There is yet another variant of the mapping functions, the modify() functions. These do not have the type variants (but the _at, _if, _depth, and so on); instead, they will always give you an output of the same type as the input:

```
modify2(1:3, 3:1, `+`)
```

```
## [1] 4 4 4
```

```
x <- c("foo", "bar", "baz")
f <- function(x, i) paste0(i, ": ", x)
x |> imodify(f)
```

```
## [1] "1: foo" "2: bar" "3: baz"
```

Reduce and Accumulate

If you want to summarize all your input into a single value, you probably want to reduce() them. Reduce repeatedly applies a function over your input sequence. If you have a function of two arguments, $f(a, x)$, and a sequence $x_1, x_2, ..., x_n$, then reduce(f) will compute $f(...f(f(x_1, x_2), x_3), ..., x_n)$, that is, it will be called on the first two elements of the sequence, the result will be paired with the next element, and so forth. Think of the argument a as an accumulator that keeps the result of the calculation so far.

To make the order of function application clear, I define a "pair" type:

```
pair <- function(first, second) {
    structure(list(first = first, second = second),
            class = "pair")
}

toString.pair <- function(x, ...) {
    first <- toString(x$first, ...)
    rest <- toString(x$second, ...)
    paste('[', first, ', ', rest, ']', sep = '')
}
print.pair <- function(x, ...) {
    x |> toString() |> cat() |> invisible()
}
```

If we reduce using pair(), we see how the values are paired when the function is called:

```
1:4 |> reduce(pair)
```

```
## [[[1, 2], 3], 4]
```

If you reverse the input, you can reduce in the opposite order, combining the last pair first and propagating the accumulator in that order.

```
1:4 |> rev() |> reduce(pair)
```

```
## [[[4, 3], 2], 1]
```

If, for some reason, you want to apply the function *and* have the accumulator as the last argument, you can use the .dir = "backward" argument.

```
1:4 |> reduce(pair, .dir = "backward")
## [1, [2, [3, 4]]]
```

The first (or last) element in the input does not have to be the value for the initial accumulator. If you want a specific starting value, you can pass that to reduce() using the .init argument.

```
1:3 |> reduce(pair, .init = 0)
## [[[0, 1], 2], 3]
1:3 |> rev() |> reduce(pair, .init = 4)
## [[[4, 3], 2], 1]
1:3 |> reduce(pair, .init = 4, .dir = "backward")
## [1, [2, [3, 4]]]
```

If your function takes more than one argument, you can provide the additional arguments to reduce() and then input sequence and function. Consider, for example, a three-argument function like this:

```
# additional arguments
loud_pair <- function(acc, next_val, volume) {
    # Build a pair
    ret <- pair(acc, next_val)
    # Announce that pair to the world
    ret |> toString() |>
        paste(volume, '\n', sep = '') |>
        cat()
    # Then return the new pair
    ret
}
```

It builds a pair object but, as a side effect, prints the pair followed by a string that indicates how "loud" the printed value is. We can provide the volume as an extra argument to reduce():

```
1:3 |>
```

```
    reduce(loud_pair, volume = '!') |>
    invisible()

    ## [1, 2]!
    ## [[1, 2], 3]!

1:3 |>
    reduce(loud_pair, volume = '!!') |>
    invisible()

## [1, 2]!!
## [[1, 2], 3]!!
```

If you want to reduce two sequences instead of one—similar to a second argument to reduce() but a sequence instead of a single value—you can use reduce2():

```
volumes <- c('!', '!!')
1:3 |> reduce2(volumes, loud_pair) |> invisible()

## [1, 2]!
## [[1, 2], 3]!!

1:3 |>
    reduce2(c('!', '!!', '!!!'), .init = 0, loud_pair) |>
    invisible()

## [0, 1]!
## [[0, 1], 2]!!
## [[[0, 1], 2], 3]!!!
```

If you want all the intermediate values of the reductions, you can use the accumulate() function. It returns a sequence of the results of each function application.

```
res <- 1:3 |> accumulate(pair)
print(res[[1]])

## [1] 1

print(res[[2]])

## [1, 2]
```

```
print(res[[3]])

## [[1, 2], 3]

res <- 1:3 |> accumulate(pair, .init = 0)
print(res[[1]])

## [1] 0

print(res[[4]])

## [[[0, 1], 2], 3]

res <- 1:3 |> accumulate(
    pair, .init = 0,
    .dir = "backward"
)
print(res[[1]])

## [1, [2, [3, 0]]]

print(res[[4]])

## [1] 0
```

The accumulate2() function works like reduce2(), except that it keeps the intermediate values like accumulate() does.

Partial Evaluation and Function Composition

When you filter, map, or reduce over sequences, you sometimes want to modify a function to match the interface of purrr's functions. If you have a function that takes too many arguments for the interface, but where you can fix some of the parameters to get the application you want, you can do what is called a *partial evaluation*. This just means that you create a new function that calls the original function with some of the parameters fixed.

For example, if you filter, you want a function that takes one input value and returns one (Boolean) output value. If you want to filter the values that are less than or greater than, say, three, you can create functions for this.

```
greater_than_three <- function(x) 3 < x
less_than_three <- function(x) x < 3

1:6 |> keep(greater_than_three)

## [1] 4 5 6

1:6 |> keep(less_than_three)

## [1] 1 2
```

The drawback of doing this is that you might need to define many such functions, even if you only use each once in your pipeline.

Using the partial() function, you can bind parameters without explicitly defining new functions. For example, to bind the first parameter to <, as in the greater_than_three() function, you can use partial():

```
1:6 |> keep(partial(`<`, 3))

## [1] 4 5 6
```

By default, you always bind the first parameter(s). To bind others, you need to name which parameters to bind. The less than operator has these parameter names:

```
`<`

## function (e1, e2) .Primitive("<")
```

so you can use this partial evaluation for less_than_three():

```
1:6 |> keep(partial(`<`, e2 = 3))

## [1] 1 2
```

Similarly, you can use partial evaluation for mapping:

```
1:6 |> map_dbl(partial(`+`, 2))

## [1] 3 4 5 6 7 8

1:6 |> map_dbl(partial(`-`, 1))

## [1] 0 -1 -2 -3 -4 -5

1:3 |> map_dbl(partial(`-`, e1 = 4))
```

```
## [1] 3 2 1
```

```
1:3 |> map_dbl(partial(`-`, e2 = 4))
```

```
## [1] -3 -2 -1
```

If you need to apply more than one function, for example

```
1:3 |>
    map_dbl(partial(`+`, 2)) |>
    map_dbl(partial(`*`, 3))
```

```
## [1] 9 12 15
```

you can also simply combine the functions. The function composition, °, works as this: $(g \circ f)(x) = g(f(x))$.

So the pipeline earlier can also be written:

```
1:3 |> map_dbl(
    compose(partial(`*`, 3), partial(`+`, 2))
)
```

```
## [1] 9 12 15
```

With `partial()` and `combine()`, you can modify functions, but using them does not exactly give you code that is easy to read. A more readable alternative is using *lambda expressions*.

Lambda Expressions

Lambda expressions are a concise syntax for defining anonymous functions, that is, functions that we do not name. The name "lambda expressions" comes from "lambda calculus,"[1] a discipline in formal logic, but in computer science, it is mostly used as a synonym for anonymous functions. In some programming languages, you cannot create anonymous functions; you need to name a function to define it. In other languages,

---

[1] The name comes from the syntax for functions that uses the Greek letter lambda, λ. A function that adds two to its argument would be written as $\lambda x.x + 2$.

you have special syntax for lambda expressions. In R, you define anonymous functions the same way that you define named functions. You always define functions the same way; you only give them a name when you assign a function definition to a variable.

If this sounds too theoretical, consider this example. When we filtered values that are even, we defined a function, `is_even()`, to use as a predicate.

```
is_even <- function(x) x %% 2 == 0
1:6 |> keep(is_even)
```

```
## [1] 2 4 6
```

We defined the function using the expression `function(x) x %% 2 == 0`, and then we assigned the function to the name `is_even`. Instead of assigning the function to a name, we could use the function definition directly in the call to `keep()`:

```
1:6 |> keep(function(x) x %% 2 == 0)
```

```
## [1] 2 4 6
```

R 4 introduced new syntax for defining functions, giving us a slightly shorter form:

```
1:6 |> keep(\(x) x %% 2 == 0)
```

```
## [1] 2 4 6
```

that I tend to use in situations such as this, but this is entirely a question of taste. There is nothing wrong with the `function(...) ...` syntax.

So, R already has lambda expressions, but the syntax can be verbose, especially before R 4 and the `\(...) ...` syntax. In `purrr`, you can use a formula as a lambda expression instead of a function. You can define an `is_even()` function using a formula like this:

```
is_even_lambda <- ~ .x %% 2 == 0
1:6 |> keep(is_even_lambda)
```

```
## [1] 2 4 6
```

or use the formula directly in your pipeline.

```
1:6 |> keep(~ .x %% 2 == 0)
```

```
## [1] 2 4 6
```

The variable .x in the expression will be interpreted as the first argument to the lambda expression.

Lambda expressions are not an approach you can use to define functions in general. They only work because purrr functions understand formulae as function specifications. You cannot write

```
is_even_lambda <- ~ .x %% 2 == 0
is_even_lambda(3)
```

This will give you an error. The error message is not telling you that you try to use a formula as a function, unfortunately, but just that it cannot find the function is_even_lambda. This is because R will look for a function when you call a variable as a function and ignore other variables with the same name. If you reassign to a variable

```
f <- function(x) 2 * x
f <- 5
f(2)
```

```
## Error in f(2): could not find function "f"
```

you are told that it cannot find a function with the name you call—not that the variable does not refer to a nonfunction. That is the error you get if you attempt to call a lambda expression outside a purrr function (or certain other Tidyverse functions).

R's rule for looking for functions can be even more confusing if there is a variable in an inner scope and a function in an outer scope:

```
f <- function(x) 2 * x
g <- function() {
    f <- 5 # not a function
    f(2) # will look for a function
}
g()
```

```
## [1] 4
```

Here, the f() function in the outer scope is called because it is a function; the variable in the inner scope is ignored.

Getting back to lambda expressions in purrr functions, you can use them as more readable versions of partial evaluation.

```
1:4 |> map_dbl(~ .x / 2)
```

```
## [1] 0.5 1.0 1.5 2.0
```

```
1:3 |> map_dbl(~ 2 + .x)
```

```
## [1] 3 4 5
```

```
1:3 |> map_dbl(~ 4 - .x)
```

```
## [1] 3 2 1
```

```
1:3 |> map_dbl(~ .x - 4)
```

```
## [1] -3 -2 -1
```

Or you can use them for more readable versions of function composition.

```
1:3 |> map_dbl(~ 3 * (.x + 2))
```

```
## [1] 9 12 15
```

If you need a lambda expression with two arguments, you can use .x and .y as the first and second arguments, respectively.

```
map2_dbl(1:3, 1:3, ~ .x + .y)
```

```
## [1] 2 4 6
```

If you need more than two arguments, you can use .n for the nth argument:

```
list(1:3, 1:3, 1:3) %>% pmap_dbl(~ .1 + .2 + .3)
```

```
## [1] 0.6 0.6 0.6
```

CHAPTER 8

Manipulating Data Frames: `dplyr`

The `dplyr` package resembles the functionality in the `purrr` package, but it is designed for manipulating data frames. It will be loaded with the `tidyverse` package, but you can also load it using

```
library(dplyr)
```

The usual way that you use `dplyr` is similar to how you use `purrr`. You string together a sequence of actions in a pipeline, with the actions separated by the `%>%` operator. The difference between the two packages is that the `purrr` functions work on sequences while the `dplyr` functions work on data frames.

This package is huge and more functionality is added in each new release, so it would be impossible for me to describe everything you can do with `dplyr`. I can only give you a flavor of the package functionality and refer you to the documentation for more information.

Selecting Columns

One of the simplest operations you can do on a data frame is selecting one or more of its columns. You can do this by indexing with $:

```
iris_df <- as_tibble(iris)
print(iris_df, n = 3)

## # A tibble: 150 × 5
##    Sepal.Length Sepal.Width Petal.Length
##           <dbl>       <dbl>        <dbl>
## 1          5.1         3.5          1.4
```

111

© Thomas Mailund 2022
T. Mailund, *R 4 Data Science Quick Reference*, https://doi.org/10.1007/978-1-4842-8780-4_8

```
## 2            4.9          3            1.4
## 3            4.7          3.2          1.3
## # . . . with 147 more rows, and 2 more variables:
## # Petal.Width <dbl>, Species <fct>
```

```
head(iris_df$Species)
```

```
## [1] setosa setosa setosa setosa setosa setosa
## Levels: setosa versicolor virginica
```

This, however, does not work well with pipelines where a data frame is flowing through from action to action. In dplyr, you have the function select() for picking out columns. We explored it in Chapter 4, but let's see a few examples more.

You can give the function one or more column names, and it will give you a data frame containing only those columns.

```
iris_df |>
    select(Sepal.Length, Species) |>
    print(n = 3)
```

```
## # A tibble: 150 × 2
##    Sepal.Length Species
##           <dbl> <fct>
## 1           5.1 setosa
## 2           4.9 setosa
## 3           4.7 setosa
## # . . . with 147 more rows
```

You can also use complements when selecting columns; just put a ! in front of the column names.

```
iris_df |>
    select(!Species) |>
    print(n = 3)
```

```
## # A tibble: 150 × 4
##    Sepal.Length Sepal.Width Petal.Length
##           <dbl>       <dbl>        <dbl>
## 1           5.1         3.5          1.4
```

```
## 2              4.9           3           1.4
## 3              4.7          3.2          1.3
## # . . . with 147 more rows, and 1 more variable:
## # Petal.Width <dbl>

iris_df |>
    select(!c(Species, Sepal.Length)) |>
    print(n = 3)

## # A tibble: 150 × 3
##   Sepal.Width Petal.Length Petal.Width
##         <dbl>        <dbl>        <dbl>
## 1         3.5          1.4          0.2
## 2         3            1.4          0.2
## 3         3.2          1.3          0.2
## # . . . with 147 more rows
```

You can also use minus to remove columns from a selection:

```
iris_df |>
    select(-Species) |>
    print(n = 3)

## # A tibble: 150 × 4
##   Sepal.Length Sepal.Width Petal.Length
##          <dbl>       <dbl>        <dbl>
## 1          5.1         3.5          1.4
## 2          4.9         3            1.4
## 3          4.7         3.2          1.3
## # . . . with 147 more rows, and 1 more variable:
## # Petal.Width <dbl>

iris_df |>
    select(-c(Species, Sepal.Length)) |>
    print(n = 3)

## # A tibble: 150 × 3
##   Sepal.Width Petal.Length Petal.Width
##         <dbl>        <dbl>        <dbl>
```

```
## 1          3.5          1.4          0.2
## 2          3            1.4          0.2
## 3          3.2          1.3          0.2
## # . . . with 147 more rows
```

but ! and - do not behave exactly the same way when you provide more than one argument to select().

With ! you ask for all columns except those you exclude, so if you write

```
iris_df |>
    select(!Species, !Sepal.Length) |>
    print(n = 3)
```

```
## # A tibble: 150 × 5
##    Sepal.Length Sepal.Width Petal.Length
##           <dbl>       <dbl>        <dbl>
## 1           5.1         3.5          1.4
## 2           4.9         3            1.4
## 3           4.7         3.2          1.3
## # . . . with 147 more rows, and 2 more variables:
## # Petal.Width <dbl>, Species <fct>
```

you first ask for all columns except Species and then for all columns except Sepal.Length, and you get the union of the two selections (which includes everything).

With minus, you ask for the columns to be removed instead, so with

```
iris_df |>
    select(-Species, -Sepal.Length) |>
    print(n = 3)
## # A tibble: 150 × 3
##    Sepal.Width Petal.Length Petal.Width
##          <dbl>        <dbl>       <dbl>
## 1          3.5          1.4         0.2
## 2          3            1.4         0.2
## 3          3.2          1.3         0.2
## # . . . with 147 more rows
```

you first remove Species and then Sepal.Length so neither are included in the final table.

You do not need to use column names. You can also use indices.

```
iris_df |>
    select(1) |>
    print(n = 3)

## # A tibble: 150 × 1
##    Sepal.Length
##           <dbl>
## 1           5.1
## 2           4.9
## 3           4.7
## # . . . with 147 more rows

iris_df |>
    select(1:3) |>
    print(n = 3)

## # A tibble: 150 × 3
##    Sepal.Length Sepal.Width Petal.Length
##           <dbl>       <dbl>        <dbl>
## 1           5.1         3.5          1.4
## 2           4.9         3            1.4
## 3           4.7         3.2          1.3
## # . . . with 147 more rows
```

This, of course, is less informative about what is being selected when you read the code. The second selection earlier shows that you can extract ranges using indices the same way as we can using column names.

```
iris_df |>
    select(Petal.Length:Species) |>
    print(n = 3)

## # A tibble: 150 × 3
##    Petal.Length Petal.Width Species
##           <dbl>       <dbl> <fct>
```

```
## 1            1.4          0.2 setosa
## 2            1.4          0.2 setosa
## 3            1.3          0.2 setosa
## # . . . with 147 more rows
```

All the other functions described in Chapter 4 work for select(), of course:

```
iris_df |>
    select(starts_with("Petal")) |>
    print(n = 3)
```

```
## # A tibble: 150 × 2
##    Petal.Length Petal.Width
##           <dbl>       <dbl>
## 1            1.4          0.2
## 2            1.4          0.2
## 3            1.3          0.2
## # . . . with 147 more rows
```

```
iris_df |>
    select(-starts_with("Petal")) |>
    print(n = 3)
```

```
## # A tibble: 150 × 3
##    Sepal.Length Sepal.Width Species
##           <dbl>       <dbl>   <fct>
## 1            5.1          3.5 setosa
## 2            4.9          3   setosa
## 3            4.7          3.2 setosa
## # . . . with 147 more rows
```

```
iris_df |>
    select(starts_with("Petal"), Species) |>
    print(n = 3)
```

```
## # A tibble: 150 × 3
##    Petal.Length Petal.Width Species
##           <dbl>       <dbl> <fct>
## 1            1.4          0.2 setosa
```

```
## 2          1.4          0.2 setosa
## 3          1.3          0.2 setosa
## # . . . with 147 more rows

iris_df |>
    select(starts_with("PETAL", ignore.case = TRUE)) |>
    print(n = 3)

## # A tibble: 150 × 2
##   Petal.Length Petal.Width
##          <dbl>       <dbl>
## 1          1.4         0.2
## 2          1.4         0.2
## 3          1.3         0.2
## # . . . with 147 more rows

iris_df |>
    select(starts_with("S")) |>
    print(n = 3)

## # A tibble: 150 × 3
##   Sepal.Length Sepal.Width Species
##          <dbl>       <dbl> <fct>
## 1          5.1         3.5 setosa
## 2          4.9         3   setosa
## 3          4.7         3.2 setosa
## # . . . with 147 more rows

iris_df |>
    select(ends_with("Length")) |>
    print(n = 3)

## # A tibble: 150 × 2
##   Sepal.Length Petal.Length
##          <dbl>       <dbl>
## 1          5.1         1.4
## 2          4.9         1.4
## 3          4.7         1.3
## # . . . with 147 more rows
```

```
iris_df |>
    select(contains("ng")) |>
    print(n = 3)
```

```
## # A tibble: 150 × 2
##    Sepal.Length Petal.Length
##           <dbl>        <dbl>
## 1           5.1          1.4
## 2           4.9          1.4
## 3           4.7          1.3
## # . . . with 147 more rows
```

You can also use select() to rename columns. If you use a named parameter, the columns you select will get the names you use for the function parameters.

```
iris_df |>
    select(sepal_length = Sepal.Length,
           sepal_width = Sepal.Width) |>
    print(n = 3)
```

```
## # A tibble: 150 × 2
##    sepal_length sepal_width
##           <dbl>        <dbl>
## 1           5.1          3.5
## 2           4.9          3
## 3           4.7          3.2
## # . . . with 147 more rows
```

Since select() removes the columns you do *not* select, it might not always be your best choice for what you need to do. A similar function to select(), where you do not lose columns, is rename(). With this function, you can rename columns and keep the remaining columns as well.

```
iris_df |>
    rename(sepal_length = Sepal.Length,
           sepal_width = Sepal.Width) |>
    print(n = 3)
```

```
## # A tibble: 150 × 5
##    sepal_length sepal_width Petal.Length
##           <dbl>       <dbl>        <dbl>
## 1          5.1         3.5          1.4
## 2          4.9         3            1.4
## 3          4.7         3.2          1.3
## # . . . with 147 more rows, and 2 more variables:
## # Petal.Width <dbl>, Species <fct>
```

Filter

While select() extracts a subset of columns, the filter() function does the same for rows.

The iris_df contains three different species. We can see this using the distinct() function. This function is also from the dplyr package, and it gives you all the unique rows from selected columns.

```
iris_df |> distinct(Species)
```

```
## # A tibble: 3 × 1
##    Species
##    <fct>
## 1 setosa
## 2 versicolor
## 3 virginica
```

We can select the rows where the species is "setosa" using filter():

```
iris_df |>
   filter(Species == "setosa") |>
   print(n = 3)
```

```
## # A tibble: 50 × 5
##    Sepal.Length Sepal.Width Petal.Length
##           <dbl>       <dbl>        <dbl>
## 1          5.1         3.5          1.4
## 2          4.9         3            1.4
```

```
## 3              4.7              3.2              1.3
## # . . . with 47 more rows, and 2 more variables:
## # Petal.Width <dbl>, Species <fct>
```

We can combine `filter()` and `select()` in a pipeline to get a subset of the columns as well as a subset of the rows.

```
iris_df |>
    filter(Species == "setosa") |>
    select(ends_with("Length"), Species) |>
    print(n = 3)
```

```
## # A tibble: 50 × 3
##    Sepal.Length Petal.Length Species
##           <dbl>        <dbl> <fct>
## 1           5.1          1.4 setosa
## 2           4.9          1.4 setosa
## 3           4.7          1.3 setosa
## # . . . with 47 more rows
```

Generally, we can string together as many `dplyr` (or `purrr`) functions as we desire in pipelines.

```
iris_df |>
    filter(Species != "setosa") |>
    distinct(Species) |>
    print(n = 3)
```

```
## # A tibble: 2 × 1
##    Species
##    <fct>
## 1 versicolor
## 2 virginica
```

We can use more than one column to filter by:

```
iris_df |>
    filter(Sepal.Length > 5, Petal.Width < 0.4) |>
    print(n = 3)
```

```
## # A tibble: 15 × 5
##   Sepal.Length Sepal.Width Petal.Length
##          <dbl> <dbl> <dbl>
## 1          5.1   3.5   1.4
## 2          5.4   3.7   1.5
## 3          5.8   4     1.2
## # . . . with 12 more rows, and 2 more variables:
## # Petal.Width <dbl>, Species <fct>
```

We can also use functions as a predicate, for example, the between() function to select numbers in a given range.

```
iris_df |>
    filter(between(Sepal.Width, 2, 2.5)) |>
    print(n = 3)
## # A tibble: 19 × 5
##   Sepal.Length Sepal.Width Petal.Length
##          <dbl>        <dbl>        <dbl>
## 1          4.5          2.3          1.3
## 2          5.5          2.3          4
## 3          4.9          2.4          3.3
## # . . . with 16 more rows, and 2 more variables:
## # Petal.Width <dbl>, Species <fct>
```

We cannot use the functions starts_with(), ends_with(), etc. that we can use with select(). This does not mean, however, that we cannot filter rows using string patterns. We just need different functions. For example, we can use str_starts() from the stringr package. We return to stringr in Chapter 9.

```
iris_df |>
    filter(str_starts(Species, "v")) |>
    print(n = 3)

## # A tibble: 100 × 5
##   Sepal.Length Sepal.Width Petal.Length
##          <dbl>        <dbl>        <dbl>
## 1          7            3.2          4.7
## 2          6.4          3.2          4.5
```

```
## 3            6.9           3.1           4.9
## # . . . with 97 more rows, and 2 more variables:
## # Petal.Width <dbl>, Species <fct>
```

```
iris_df |>
    filter(str_ends(Species, "r")) |>
    print(n = 3)
```

```
## # A tibble: 50 × 5
##    Sepal.Length Sepal.Width Petal.Length
##           <dbl>       <dbl>        <dbl>
## 1            7          3.2          4.7
## 2          6.4          3.2          4.5
## 3          6.9          3.1          4.9
## # . . . with 47 more rows, and 2 more variables:
## # Petal.Width <dbl>, Species <fct>
```

While `filter()` selects rows by applying predicates on individual columns, other filter variants will use predicates over more than one column.

The `filter_all()` function will apply the predicate over all columns. You must provide an expression where you use . (dot) for the table cell value and wrap the expression in one of two functions: `any_var()` or `all_var()`. If you use `any_var()`, it suffices that one column satisfies the predicate for the row to be included; if you use `all_var()`, then all columns must satisfy the predicate.

We can require that any value in the `iris_df` data must be larger than five:

```
iris_df |>
    select(-Species) |>
    filter_all(any_vars(. > 5)) |>
    print(n = 3)
```

```
## # A tibble: 118 × 4
##    Sepal.Length Sepal.Width Petal.Length
##           <dbl>       <dbl>        <dbl>
## 1           5.1         3.5          1.4
## 2           5.4         3.9          1.7
## 3           5.4         3.7          1.5
## # . . . with 115 more rows, and 1 more variable:
```

```
## # Petal.Width <dbl>
```

The expression . > 5 is not meaningful if . is a string, so I had to remove the Species column before I filtered. Rarely will we accept to lose an informative variable. If you want to keep a column, but not apply the predicate to values in it, you can use the filter_at() function.

```
iris_df |>
    filter_at(vars(-Species), any_vars(. > 5)) |>
    print(n = 3)
```

```
## # A tibble: 118 × 5
##   Sepal.Length Sepal.Width Petal.Length
##          <dbl>       <dbl>        <dbl>
## 1          5.1         3.5          1.4
## 2          5.4         3.9          1.7
## 3          5.4         3.7          1.5
## # . . . with 115 more rows, and 2 more variables:
## # Petal.Width <dbl>, Species <fct>
```

You can give filter_at() a vector of strings or a vector of variables (unquoted column names). Because I need to negate Species to select all other columns, I used vars() here. In other cases, you can use either option.

```
iris_df |>
    filter_at(c("Petal.Length", "Sepal.Length"),
    any_vars(. > 0)) |>
    print(n = 3)
```

```
## # A tibble: 150 × 5
##   Sepal.Length Sepal.Width Petal.Length
##          <dbl>       <dbl>        <dbl>
## 1          5.1         3.5          1.4
## 2          4.9         3            1.4
## 3          4.7         3.2          1.3
## # . . . with 147 more rows, and 2 more variables:
## # Petal.Width <dbl>, Species <fct>
```

```
iris_df |>
    filter_at(vars(Petal.Length, Sepal.Length),
    any_vars(. > 0)) |>
    print(n = 3)
```

```
## # A tibble: 150 × 5
##    Sepal.Length Sepal.Width Petal.Length
##           <dbl>       <dbl>        <dbl>
## 1           5.1         3.5          1.4
## 2           4.9         3            1.4
## 3           4.7         3.2          1.3
## # . . . with 147 more rows, and 2 more variables:
## # Petal.Width <dbl>, Species <fct>
```

You can use `filter_if()` to use a predicate over the columns:

```
iris_df |>
    filter_if(is.numeric, all_vars(. < 5)) |>
    print(n = 3)
```

```
## # A tibble: 22 × 5
##    Sepal.Length Sepal.Width Petal.Length
##           <dbl>       <dbl>        <dbl>
## 1           4.9         3            1.4
## 2           4.7         3.2          1.3
## 3           4.6         3.1          1.5
## # . . . with 19 more rows, and 2 more variables:
## # Petal.Width <dbl>, Species <fct>
```

You can use functions or lambda expressions as the predicate.

Consider a case where we have numbers with missing data in a table. If we want to filter the rows where all values are greater than three, we can use `filter_all()`:

```
df <- tribble(
    ~A, ~B, ~C,
     1,  2,  3,
     4,  5, NA,
    11, 12, 13,
```

```
   22, 22,   1
)
df |> filter_all(all_vars(. > 3))
```

```
## # A tibble: 1 × 3
##       A     B     C
##   <dbl> <dbl> <dbl>
## 1    11    12    13
```

It removes the first two and the last row because they all contain values smaller or equal to three. It also deletes the third row because NA is not considered greater than three. We can restrict the tests to the columns that do not contain missing values:

```
df |> filter_if(~ all(!is.na(.)), all_vars(. > 3))
```

```
## # A tibble: 3 × 3
##       A     B     C
##   <dbl> <dbl> <dbl>
## 1     4     5    NA
## 2    11    12    13
## 3    22    22     1
```

This removes column C from the tests. It means that we also keep the last row even though C has a value smaller than three.

If you want to keep the two middle row but not the first or last, you can write a more complex predicate and use filter_all().

```
df |> filter_all(all_vars(is.na(.) | . > 3))
```

```
## # A tibble: 2 × 3
##       A     B     C
##   <dbl> <dbl> <dbl>
## 1     4     5    NA
## 2    11    12    13
```

You need to use the | vector-or. The expression needs to be a vector expression so you cannot use ||.

Many dplyr functions have _all, _at, and _if variants, but I will only describe the frequently used functions here. For the others, I refer to the package documentation.

Sorting

If you want to sort rows by values in selected columns, the function you want is called arrange(). You can sort by one or more columns:

```
iris_df |>
    arrange(Petal.Length) |>
    print(n = 5)
## # A tibble: 150 × 5
##   Sepal.Length Sepal.Width Petal.Length
##          <dbl>       <dbl>        <dbl>
## 1          4.6         3.6            1
## 2          4.3         3            1.1
## 3          5.8         4            1.2
## 4          5           3.2          1.2
## 5          4.7         3.2          1.3
## # . . . with 145 more rows, and 2 more variables:
## # Petal.Width <dbl>, Species <fct>

iris_df |>
    arrange(Sepal.Length, Petal.Length) |>
    print(n = 5)

## # A tibble: 150 × 5
##   Sepal.Length Sepal.Width Petal.Length
##          <dbl>       <dbl>        <dbl>
## 1          4.3         3            1.1
## 2          4.4         3            1.3
## 3          4.4         3.2          1.3
## 4          4.4         2.9          1.4
## 5          4.5         2.3          1.3
## # . . . with 145 more rows, and 2 more variables:
## # Petal.Width <dbl>, Species <fct>
```

If you want to sort in descending order, you can use the function desc().

```
iris_df |>
    arrange(desc(Petal.Length)) |>
```

```
   print(n = 5)
```

```
## # A tibble: 150 × 5
##   Sepal.Length Sepal.Width Petal.Length
##          <dbl>       <dbl>        <dbl>
## 1         7.7         2.6          6.9
## 2         7.7         3.8          6.7
## 3         7.7         2.8          6.7
## 4         7.6         3            6.6
## 5         7.9         3.8          6.4
## # . . . with 145 more rows, and 2 more variables:
## # Petal.Width <dbl>, Species <fct>
```

Modifying Data Frames

When we work with data frames, we usually want to compute values based on the variables (columns) in our tables. Filtering rows and columns will only get us so far.

The mutate() function lets us add columns to a data frame based on expressions that can involve any of the existing columns. Consider a table of widths and heights.

```
df <- tribble(
    ~height, ~width,
        10,     12,
        42,     24,
        14,     12
)
```

We can add an area column using mutate() and this expression:

```
df |> mutate(area = height * width)
```

```
## # A tibble: 3 × 3
##   height width  area
##    <dbl> <dbl> <dbl>
## 1     10    12   120
## 2     42    24  1008
## 3     14    12   168
```

If you add columns to a data frame, you can refer to variables you have already added. For example, if your height and width data are in inches and you want them in centimeters, and you also want the area in centimeters squared, you can do this:

```
cm_per_inch <- 2.54
df |> mutate(
    height_cm = cm_per_inch * height,
    width_cm = cm_per_inch * width,
    area_cm = height_cm * width_cm
)
```

```
## # A tibble: 3 × 5
##    height width height_cm width_cm area_cm
##     <dbl> <dbl>     <dbl>    <dbl>   <dbl>
## 1      10    12      25.4     30.5    774.
## 2      42    24      107.     61.0   6503.
## 3      14    12      35.6     30.5   1084.
```

The following expression, however, will not work since you cannot refer to variables you are adding later in the mutate() call:

```
df %>% mutate(
    area_cm = height_cm * width_cm,
    height_cm = cm_per_inch * height,
    width_cm = cm_per_inch * width
)
```

```
## Error in `mutate()`:
## ! Problem while computing `area_cm = height_cm * width_cm`.
## Caused by error:
## ! object 'height_cm' not found
```

If you do not give the new variable a name, that is, you call mutate() with named parameters, then the expression will become the variable name:

```
df |> mutate(cm_per_inch * height)
```

```
## # A tibble: 3 × 3
##    height width `cm_per_inch * height`
```

```
##      <dbl> <dbl>            <dbl>
## 1      10    12             25.4
## 2      42    24            107.
## 3      14    12             35.6
```

In this example, the units (inches and centimeters) are not encoded with the data, and in future computations you might end up mixing the wrong units. If you use the units package—which is not part of the Tidyverse—you can make the units explicit, automatically convert between units where that makes sense, and you will get unit type-safety as a bonus.

```
library(units)
df |> mutate(
    height_in = units::as_units(height, "in"),
    width_in = units::as_units(width, "in"),
    area_in = height_in * width_in,
    height_cm = units::set_units(height_in, "cm"),
    width_cm = units::set_units(width_in, "cm"),
    area_cm = units::set_units(area_in, "cm^2")
)
```

```
## # A tibble: 3 × 8
##   height width height_in width_in area_in
##    <dbl> <dbl>      [in]     [in] [in^2]
## 1     10    12        10       12    120
## 2     42    24        42       24   1008
## 3     14    12        14       12    168
## # . . . with 3 more variables: height_cm [cm],
## # width_cm [cm], area_cm [cm^2]
```

This is possible because tibbles can contain classes of various types, which the units package exploits. Tibbles use an interface that lets you define how your own classes should be displayed,[1] but this is beyond the scope of this book.

---

[1] http://tinyurl.com/yya2btby

Earlier, we created unit-carrying values for the original data table, but the original height and weight columns are still around. We could use select() to remove them, but the function transmute() already does this for us.

```
df |> transmute(
    height_in = units::as_units(height, "in"),
    width_in = units::as_units(width, "in"),
    area_in = height_in * width_in,
    height_cm = units::set_units(height_in, "cm"),
    width_cm = units::set_units(width_in, "cm"),
    area_cm = units::set_units(area_in, "cm^2")
)

## # A tibble: 3 × 6
##    height_in width_in area_in height_cm width_cm
##         [in]     [in]  [in^2]      [cm]     [cm]
## 1         10       12     120      25.4     30.5
## 2         42       24    1008      107.     61.0
## 3         14       12     168      35.6     30.5
## # . . . with 1 more variable: area_cm [cm^2]
```

When you use transmute() instead of mutate(), the variables that you are not assigning to will be removed.

As with filtering, the expressions you use when mutating a data frame must be vector expressions. Consider this example:

```
df <- tibble(
    x = rnorm(3, mean = 12, sd = 5),
)
df |> mutate(~ if (.x < 0) -.x else .x)
## Error in `mutate()`:
## ! Problem while computing `..1 = ~if (.x < 0) -.x else .x`.
## x `..1` must be a vector, not a `formula` object.
```

It fails because the function my_abs() (my version of the abs() function) is not a vector expression. It is not a vector expression because of the x in the if expression.

If we use the built-in abs() function, then there is no problem with the expression; it handles vectors.

```
df |> mutate(abs(x))
```

```
## # A tibble: 3 × 2
##       x `abs(x)`
##   <dbl> <dbl>
## 1  11.5  11.5
## 2  20.9  20.9
## 3  17.2  17.2
```

If you use the `ifelse()` vector version of `if` expressions, you get a vector expression.

```
df |> mutate(ifelse(x < 0, -x, x))
```

```
## # A tibble: 3 × 2
##       x `ifelse(x < 0, -x, x)`
##   <dbl> <dbl>
## 1  11.5  11.5
## 2  20.9  20.9
## 3  17.2  17.2
```

Or you can use the `Vectorize()` function to make a vector expression out of a function that does not handle vectors.

```
my_abs <- Vectorize(\(x) if (x < 0) -x else x)
df |> mutate(my_abs(x))
```

```
## # A tibble: 3 × 2
##       x `my_abs(x)`
##   <dbl>     <dbl>
## 1  11.5      11.5
## 2  20.9      20.9
## 3  17.2      17.2
```

If you need to map the input to several different output values, you can nest `ifelse()` expressions arbitrarily deep, but it gets difficult to read. A function that alleviates this problem substantially is `case_when()`. You can give it a sequence of predicates with matching expressions, and it will return the expression for the first matching predicate.

Consider this example where we use `case_when()` to categorize the variable x based on its value:

```
df <- tibble(x = rnorm(100))
df |>
    mutate(
        x_category = case_when(
            x - mean(x) < -2 * sd(x) ~ "small",
            x - mean(x) >  2 * sd(x) ~ "large",
            TRUE                     ~ "medium"
        )
    ) |>
    print(n = 3)

## # A tibble: 100 × 2
##         x x_category
##     <dbl> <chr>
## 1  -0.736 medium
## 2  -0.478 medium
## 3   0.796 medium
## # . . . with 97 more rows
```

The TRUE line corresponds to the else part of an if-else statement.

Grouping and Summarizing

In many analyses, we need summary statistics of our data. If you can map one or more of your columns into a single summary, you can use the `summarise()` function.

```
df <- tibble(x = rnorm(100), y = rnorm(100))
df |> summarise(mean_x = mean(x), mean_y = mean(y))

## # A tibble: 1 × 2
##   mean_x mean_y
##    <dbl>  <dbl>
## 1 -0.154 0.0224
```

In this example, we summarized our data as the mean of variables x and y.

If you split your data into different classes and want to work on the data per group, you can use the function group_by().

```
classify <- function(x) {
    case_when(
        x - mean(x) < -2 * sd(x) ~ "small",
        x - mean(x) >  2 * sd(x) ~ "large",
        TRUE                     ~ "medium"
    )
}
df |> mutate(x_category = classify(x)) |>
    group_by(x_category) |>
    print(n = 3)

## # A tibble: 100 × 3
## # Groups:    x_category [3]
##        x      y x_category
##    <dbl>  <dbl> <chr>
## 1  -2.08 -1.52  medium
## 2  0.580 -0.990 medium
## 3 -0.995 -0.443 medium
## # . . . with 97 more rows
```

The result is a data frame that, when we print it, doesn't look different from before we grouped the data. In the header, however, you will notice the line:

```
# Groups:    x_category [3]
```

This tells you that something *is* different when you have grouped your data. This is apparent when you combine grouping with summarizing.

```
df |> mutate(x_category = classify(x)) |>
    group_by(x_category) |>
    summarise(mean_x = mean(x), no_x = n())

## # A tibble: 3 × 3
##   x_category mean_x   no_x
##   <chr>       <dbl>  <int>
## 1 large        2.04      2
```

```
## 2 medium       -0.155      96
## 3 small         -2.30       2
```

When you group your data and then summarize, you get a per-group summary statistics. Here, we calculate the mean and the number of observations in each category (the function n() gives you the number of observations).

You can get the variables your data is grouped by using the group_vars() function.

```
df |> mutate(x_category = classify(x)) |>
    group_by(x_category) |>
    group_vars()
```

```
## [1] "x_category"
```

You can group data by more than one variable.

```
df <- tibble(x = rnorm(100), y = rnorm(100)) |>
    mutate(
      x_category = classify(x),
      y_category = classify(y)
    )
```

```
df |> group_by(x_category, y_category) |>
    group_vars()
```

```
## [1] "x_category" "y_category"
```

If you do, you will get summaries for each combination of the grouping variables.

```
df |> group_by(x_category, y_category) |>
    summarise(mean_x = mean(x), mean_y = mean(y))
```

```
## `summarise()` has grouped output by 'x_category'. You can override
using the
## `.groups` argument.
```

```
## # A tibble: 5 × 4
## # Groups: x_category [3]
##    x_category y_category mean_x mean_y
##    <chr>      <chr>       <dbl>  <dbl>
## 1 large      medium      2.08    1.33
```

```
## 2 medium    large     0.808    2.69
## 3 medium    medium    0.0755   0.141
## 4 medium    small     1.05    -2.38
## 5 small     medium   -2.55    -1.12
```

The notice you get that the output has "grouped output by 'x_category'" means just that: the table we get out from the call to summarise() is still grouped by x_category, as you can also see in the printed output of the result (in the Groups: x_category [3] line in the header).

When you group by multiple variables, it works in effect as if you add groups one by one, and you can also do this via multiple calls to group_by(). There, however, you need the argument .add = TRUE to add a group rather than replacing one, which is what group_by() will do without this argument.

```
# Adding two categories at the same time
df |> group_by(x_category, y_category) |> group_vars()
```

```
## [1] "x_category" "y_category"
```

```
# Adding categories in two steps
df |> group_by(x_category) |> group_by(y_category, .add = TRUE) |>
group_vars()
```

```
## [1] "x_category" "y_category"
```

```
# Replacing one grouping with another
df |> group_by(x_category) |> group_by(y_category) |> group_vars()
```

```
## [1] "y_category"
```

When you summarize, you remove the last grouping you created—the summary only has one row per value in that group anyway, so it isn't useful any longer—but the other groups remain. This was the default behavior in early versions of dplyr, and still is, but is a common source of errors. Some data scientists expect this behavior, others expect that all groups are removed, and if your expectations are not met, there is a good chance that the following analysis will be wrong; if you think you are working on the data as a whole, but all operations are grouped by one or more variables, you are unlikely to get the results that you want.

Because of this, `summarise()` now takes the argument `.groups` where you can specify the behavior you want. It is not yet required, since that would break backward compatibility to a lot of code, but you get the message about using it that we saw a little while up.

```
# Drop the last group we made
df |> group_by(x_category, y_category) |>
      summarise(mean_x = mean(x), mean_y = mean(y), .groups = "drop_last") |>
      group_vars()
```

```
## [1] "x_category"
# Drop all groups
df |> group_by(x_category, y_category) |>
      summarise(mean_x = mean(x), mean_y = mean(y), .groups = "drop") |>
      group_vars()
```

```
## character(0)
```

```
# Keep all groups
df |> group_by(x_category, y_category) |>
      summarise(mean_x = mean(x), mean_y = mean(y), .groups = "keep") |>
      group_vars()
```

```
## [1] "x_category" "y_category"
```

If you have grouped your data, or you get grouped data from one summary as before, you can remove the groups again using `ungroup()`. Once you have done this, you can compute global summaries again.

```
df |> group_by(x_category, y_category) |>
      ungroup() |> # Remove the grouping again
      group_vars()
```

```
## character(0)
```

Grouping is not only useful when collecting summaries of your data. You can also add columns to your data frame based on per-group computations using `mutate()`:

```
df |> group_by(x_category) |>
      mutate(mean_x = mean(x), mean_y = mean(y)) |>
```

```
      ungroup() |>
      print(n = 5)
```

```
## # A tibble: 100 × 6
##          x          y x_category y_category mean_x
##      <dbl>      <dbl> <chr>      <chr>       <dbl>
## 1 -0.416    -0.374    medium     medium      0.123
## 2 -0.845    -1.21     medium     medium      0.123
## 3 -0.0176    0.103    medium     medium      0.123
## 4 -1.18     -0.0132   medium     medium      0.123
## 5  0.508     0.504    medium     medium      0.123
## # . . . with 95 more rows, and 1 more variable:
## #   mean_y <dbl>
```

When you calculate a summary and then mutate, you create columns where a summary parameter takes values based on the groupings, but they are combined with the existing data rather than extracted as a new data frame as summarise() would do.

A mutate() without a group_by() will give you summaries for the entire data.

```
df |> mutate(mean_x = mean(x), mean_y = mean(y)) |>  # Calculate mean x and
                                                     # y for entire data
      distinct(mean_x, mean_y)                       # Get unique values for
                                                     # printing
```

```
## # A tibble: 1 × 2
##   mean_x mean_y
##    <dbl>  <dbl>
## 1 0.0895 0.0520
```

I have used distinct() to get the unique values of r mean_x and mean_y combinations, as this well illustrate the result. Here, we only have one row of values, and this is because we have computed the mean values for the entire data set and not for different groupings of the data.

In contrast, if you group before adding variables to the data, then the summaries are per group.

```
df |> group_by(x_category) |>                        # Group by x
                                                     # categories only
```

```
    mutate(mean_x = mean(x), mean_y = mean(y)) |> # Calculate mean x and
                                                    y for each x group
    distinct(mean_x, mean_y)                      # Get unique values for
                                                    printing
```

```
## # A tibble: 3 × 3
## # Groups:   x_category [3]
##   x_category mean_x  mean_y
##   <chr>       <dbl>   <dbl>
## 1 medium      0.123  0.0631
## 2 large       2.08   1.33
## 3 small      -2.55  -1.12
```

Here, there are values for different x_category factors, showing us that we have computed means for each different x_category in the data.

You can combine data-wide summaries with grouped summaries if you calculate the global summaries before you group.

```
df |> mutate(mean_y = mean(y)) |> # Get the mean for y over all data points
    group_by(x_category)      |> # Then group so the following works
                                   per group
    mutate(mean_x = mean(x)) |> # Get the mean x for each group (not
                                   globally)
    distinct(mean_x, mean_y)     # Get the unique values so we can print
                                   the result
```

```
## # A tibble: 3 × 3
## # Groups:   x_category [3]
##   x_category mean_y  mean_x
##   <chr>       <dbl>   <dbl>
## 1 medium     0.0520  0.123
## 2 large      0.0520  2.08
## 3 small      0.0520 -2.55
```

Notice that the mean_y values are the same in the output rows; this is because we have computed the mean globally and not for each group.

If you need to compute summaries with different groups, for example, the mean of x for each x_category as well as the mean of y for each y_category, then you can call group_by(x_category) for summarizing x, followed by group_by(y_category) for changing the grouping so you can summarize y.

```
df |> group_by(x_category) |>
    mutate(mean_x = mean(x)) |>
    group_by(y_category) |>
    mutate(mean_y = mean(y)) |>
    distinct(
      x_category, mean_x,
      y_category, mean_y
    )
```

```
## # A tibble: 5 × 4
## # Groups:  y_category [3]
##    x_category y_category mean_x mean_y
##    <chr>      <chr>       <dbl>  <dbl>
## 1 medium     medium      0.123   0.127
## 2 medium     small       0.123  -2.38
## 3 large      medium      2.08    0.127
## 4 medium     large       0.123   2.69
## 5 small      medium     -2.55    0.127
```

Joining Tables

It is not uncommon to have your data in more than one table. This could be because the tables are created from different calculations, for example, different kinds of summaries, or it can be because you got the data from different files.

If you merely need to combine tables by row or column, then you can use the bind_rows() and bind_columns() functions which do precisely what you would expect.

```
df1 <- tibble(
    A = paste0("a", 1:2),
    B = paste0("b", 1:2)
)
```

```r
df2 <- tibble(
    A = paste0("a", 3:4),
    B = paste0("b", 3:4)
)
df3 <- tibble(
    C = paste0("c", 1:2),
    D = paste0("d", 1:2)
)
bind_rows(df1, df2)
```

```
## # A tibble: 4 × 2
##    A     B
##    <chr> <chr>
## 1 a1    b1
## 2 a2    b2
## 3 a3    b3
## 4 a4    b4
```

```r
bind_cols(df1, df3)
```

```
## # A tibble: 2 × 4
##    A     B     C     D
##    <chr> <chr> <chr> <chr>
## 1 a1    b1    c1    d1
## 2 a2    b2    c2    d2
```

When you combine rows, the tables must have the same columns, and when you combine by column, the tables must have the same number of rows.

If you have tables that represent different relations between variables—the underlying principle of relational databases aimed at avoiding duplicated data—then you can combine them using join functions.

Say you have a table that maps students to grades for each class you teach. You can join tables from two different classes using inner_join(). You use a key to join them on, specified by argument by.

```r
grades_maths <- tribble(
    ~name, ~grade,
```

```
    "Marko Polo", "D",
    "Isaac Newton", "A+",
    "Charles Darwin", "B"
)
grades_biology <- tribble(
    ~name, ~grade,
    "Marko Polo", "F",
    "Isaac Newton", "D",
    "Charles Darwin", "A+"
)

inner_join(grades_maths, grades_biology, by = "name")
## # A tibble: 3 × 3
##    name          grade.x grade.y
##    <chr>         <chr>   <chr>
## 1 Marko Polo     D       F
## 2 Isaac Newton   A+      D
## 3 Charles Darwin B       A+
```

This tells inner_join() that you want to combine all rows in the first table with all rows in the second, where the two rows have the same name. You can use more than one key in a join if you give by a vector of variable names.

Earlier, each name appears once per table. If a key appears more than once, then the result of an inner join will have a list with all combinations of rows sharing a name.

```
grades_maths2 <- tribble(
    ~name, ~grade,
    "Marko Polo", "D",
    "Isaac Newton", "A+", # so good at physics
    "Isaac Newton", "A+", # that he got an A+ twice
    "Charles Darwin", "B"
)
grades_biology2 <- tribble(
    ~name, ~grade,
    "Marko Polo", "F",
    "Isaac Newton", "D",
```

```
    "Charles Darwin", "A+", # so good at biology that we
    "Charles Darwin", "A+" # listed him twice
)
inner_join(grades_maths2, grades_biology2, by = "name")
## # A tibble: 5 × 3
##    name            grade.x grade.y
##    <chr>           <chr>   <chr>
## 1 Marko Polo       D       F
## 2 Isaac Newton     A+      D
## 3 Isaac Newton     A+      D
## 4 Charles Darwin   B       A+
## 5 Charles Darwin   B       A+

inner_join(grades_maths2, grades_biology2, by = "grade")

## # A tibble: 5 × 3
##    name.x          grade name.y
##    <chr>           <chr> <chr>
## 1 Marko Polo       D     Isaac Newton
## 2 Isaac Newton     A+    Charles Darwin
## 3 Isaac Newton     A+    Charles Darwin
## 4 Isaac Newton     A+    Charles Darwin
## 5 Isaac Newton     A+    Charles Darwin
```

In the last join, you see that you can get the same line multiple times from an inner join. Combine the join with distinct() if you want to avoid this.

```
inner_join(grades_maths2, grades_biology2, by = "grade") |> distinct()

## # A tibble: 2 × 3
##    name.x          grade name.y
##    <chr>           <chr> <chr>
## 1 Marko Polo       D     Isaac Newton
## 2 Isaac Newton     A+    Charles Darwin
```

The tables can have different variables—if they represent different relations—or they can share several or all variables. In the grade example, each table contains the same name/grade relationships, just for different classes. Therefore, they also share variables.

With shared variables, the data from the two tables can be differentiated by a suffix. The default, see above, is ".x" and ".y". You can change that using the suffix argument.

```
inner_join(
    grades_maths, grades_biology,
    by = "name", suffix = c(".maths", ".biology")
)
```

```
## # A tibble: 3 × 3
##   name           grade.maths grade.biology
##   <chr>          <chr>       <chr>
## 1 Marko Polo     D           F
## 2 Isaac Newton   A+          D
## 3 Charles Darwin B           A+
```

Students might take different classes, and you have several choices on how to combine tables with different keys.

An inner_join() will only give you the rows where a key is in both tables.

```
grades_geography <- tribble(
    ~name, ~grade,
    "Marko Polo", "A",
    "Charles Darwin", "A",
    "Immanuel Kant", "A+"
)
grades_physics <- tribble(
    ~name, ~grade,
    "Isaac Newton", "A+",
    "Albert Einstein", "A+",
    "Charles Darwin", "C"
)
```

```
inner_join(
    grades_geography, grades_physics,
    by = "name", suffix = c(".geography", ".physics")
)
```

```
## # A tibble: 1 × 3
##    name           grade.geography grade.physics
##    <chr>          <chr>           <chr>
## 1 Charles Darwin A               C
```

The full_join() function, in contrast, gives you a table containing all keys. If a key is only found in one of the tables, the variables from the other table will be set to NA.

```
full_join(
    grades_geography, grades_physics,
    by = "name", suffix = c(".geography", ".physics")
)
```

```
## # A tibble: 5 × 3
##    name            grade.geography grade.physics
##    <chr>           <chr>           <chr>
## 1 Marko Polo      A               <NA>
## 2 Charles Darwin  A               C
## 3 Immanuel Kant   A+              <NA>
## 4 Isaac Newton    <NA>            A+
## 5 Albert Einstein <NA>            A+
```

If you want all keys from the left or right table (but not both left and right)—potentially with NA if the other table does not have the key—then you need left_join() or right_join().

```
left_join(
    grades_geography, grades_physics,
    by = "name", suffix = c(".geography", ".physics")
)
```

```
## # A tibble: 3 × 3
##    name           grade.geography grade.physics
##    <chr>          <chr>           <chr>
## 1 Marko Polo     A               <NA>
## 2 Charles Darwin A               C
## 3 Immanuel Kant  A+              <NA>
```

```
right_join(
    grades_maths, grades_physics,
    by = "name", suffix = c(".maths", ".physics")
)
```

```
## # A tibble: 3 × 3
##   name            grade.maths grade.physics
##   <chr>           <chr>       <chr>
## 1 Isaac Newton    A+          A+
## 2 Charles Darwin  B           C
## 3 Albert Einstein <NA>        A+
```

A semi_join() will give you all the rows in the first table that contains a key in the second table.

```
semi_join(
    grades_maths2, grades_biology2,
    by = "name", suffix = c(".geography", ".physics")
)
```

```
## # A tibble: 4 × 2
##   name           grade
##   <chr>          <chr>
## 1 Marko Polo     D
## 2 Isaac Newton   A+
## 3 Isaac Newton   A+
## 4 Charles Darwin B
```

You only get one copy per row in the first table when a key matches, regardless of how many times the key appears in the second table.

This is unlike inner_join() where you get all combinations of rows from the two tables. The inner_join() is, therefore, not an analogue to an inner_join() combined with a select().

```
inner_join(
    grades_maths2, grades_biology2,
    by = "name", suffix = c(".geography", ".physics")
) |> select(1:2)
```

```
## # A tibble: 5 × 2
##   name           grade.geography
##   <chr>          <chr>
## 1 Marko Polo     D
## 2 Isaac Newton   A+
## 3 Isaac Newton   A+
## 4 Charles Darwin B
## 5 Charles Darwin B
```

You can still get multiple identical rows from a `semi_join()`. If the first table has duplicated rows, you will get the same duplication of the rows. If you do not want that, you can combine the join with `distinct()`.

An `anti_join()` gives you all the rows in the first table where the key or keys are not found in the second table. Think of it as the complement of `semi_join()`.

```
anti_join(
    grades_maths2, grades_physics,
    by = "name", suffix = c(".geography", ".physics")
)
```

```
## # A tibble: 1 × 2
##   name        grade
##   <chr>       <chr>
## 1 Marko Polo  D
```

The join functions only take two tables as input so you might wonder how you can combine multiple tables. One solution is to use purrr's `reduce()` function:

```
grades <- list(
    grades_maths, grades_biology,
    grades_geography, grades_physics
)
grades |>
    reduce(full_join, by = "name") |>
    rename_at(2:5, ~ c("maths", "biology", "geography", "physics"))
```

```
## # A tibble: 5 × 5
##   name          maths biology geography physics
##   <chr>         <chr> <chr>   <chr>     <chr>
```

```
## 1 Marko Polo       D      F      A        <NA>
## 2 Isaac Newton     A+     D      <NA>     A+
## 3 Charles Darwin   B      A+     A        C
## 4 Immanuel Kant    <NA>   <NA>   A+       <NA>
## 5 Albert Einstein  <NA>   <NA>   <NA>     A+
```

The rename_at() function works similarly to select_at(), and here, I use it to rename the last four columns.

Income in Fictional Countries

We had an example with income in fictional countries in Chapter 5. There, we did not have the proper tools needed to deal with missing data. We wanted to replace NA with the mean income for a country in rows with missing data, but the functions from tidyr didn't suffice. With dplyr we have the tools.

Recall the data:

```
mean_income <- tribble(
    ~country, ~`2002`, ~`2003`, ~`2004`, ~`2005`,
    "Numenor",  123456, 132654,     NA, 324156,
    "Westeros", 314256,     NA,     NA, 465321,
    "Narnia",   432156,     NA,     NA,     NA,
    "Gondor",   531426, 321465, 235461, 463521,
    "Laputa",    14235,  34125,  45123,  51234,
)
```

The following pipeline does what we want. I will explain it as follows, but you can try to work out the steps:

```
mean_income                                              |>
    pivot_longer(
        names_to = "year",
        values_to = "mean_income",
        !country
    )                                                    |>
    group_by(
        country
```

```
    )                                                              |>
    mutate(
        mean_per_country = mean(mean_income, na.rm = TRUE),
        mean_income = ifelse(
            is.na(mean_income),
            mean_per_country,
            mean_income
        )
    )                                                              |>
    pivot_wider(
        names_from = "year",
        values_from = "mean_income"
    )
```

```
## # A tibble: 5 × 6
## # Groups: country [5]
##   country mean_per_country `2002` `2003` `2004`
##   <chr>              <dbl>  <dbl>  <dbl>  <dbl>
## 1 Numenor           193422 123456 132654 193422
## 2 Westeros          389788. 314256 389788. 389788.
## 3 Narnia            432156 432156 432156 432156
## 4 Gondor            387968. 531426 321465 235461
## 5 Laputa             36179.  14235  34125  45123
## # . . . with 1 more variable: `2005` <dbl>
```

The first step in the pipeline is reformatting the data. We saw how this works in Chapter 5.

```
mean_income |>
    pivot_longer(
        names_to = "year",
        values_to = "mean_income",
        !country
    )
```

```
## # A tibble: 20 × 3
##   country   year   mean_income
```

```
##    <chr>     <chr>         <dbl>
## 1 Numenor   2002         123456
## 2 Numenor   2003         132654
## 3 Numenor   2004             NA
## 4 Numenor   2005         324156
## 5 Westeros  2002         314256
## 6 Westeros  2003             NA
## 7 Westeros  2004             NA
## 8 Westeros  2005         465321
## 9 Narnia    2002         432156
## 10 Narnia   2003             NA
## 11 Narnia   2004             NA
## 12 Narnia   2005             NA
## 13 Gondor   2002         531426
## 14 Gondor   2003         321465
## 15 Gondor   2004         235461
## 16 Gondor   2005         463521
## 17 Laputa   2002          14235
## 18 Laputa   2003          34125
## 19 Laputa   2004          45123
## 20 Laputa   2005          51234
```

We use pivot_longer() to get all the income data from separate columns into a single mean_income column, where the year column will hold the information about which original column the data came from.

In the next two steps, we group by countries, so the means we calculate are per country and not the full data set, and then we compute the means and replace NA cells with them. To get a glimpse into the mutate() call, you can replace it with summarise() and see the mean of each country.

```
mean_income |>
    pivot_longer(
        names_to = "year",
        values_to = "mean_income",
        !country
    )                                                        |>
```

```
    group_by(
        country
    )                                                      |>
    summarise(
        per_country_mean = mean(mean_income, na.rm = TRUE)
    )
```

```
## # A tibble: 5 × 2
##    country  per_country_mean
##    <chr>               <dbl>
## 1 Gondor            387968.
## 2 Laputa             36179.
## 3 Narnia            432156
## 4 Numenor           193422
## 5 Westeros          389788.
```

In the next step, we can replace the missing data with their country mean in the
mutate() call because we can refer to earlier columns when we create later columns in
mutate().

```
mean_income                                                |>
    pivot_longer(
        names_to = "year",
        values_to = "mean_income",
        !country
    )                                                      |>
    group_by(
        country
    )                                                      |>
    mutate(
        mean_per_country = mean(mean_income, na.rm = TRUE),
        mean_income = ifelse(
            is.na(mean_income),
            mean_per_country,
            mean_income
        )
    )
```

```
## # A tibble: 20 × 4
## # Groups:   country [5]
##    country year  mean_income mean_per_country
##    <chr>   <chr>       <dbl>            <dbl>
## 1  Numenor 2002       123456           193422
## 2  Numenor 2003       132654           193422
## 3  Numenor 2004       193422           193422
## 4  Numenor 2005       324156           193422
## 5  Westeros 2002      314256           389788.
## 6  Westeros 2003      389788.          389788.
## 7  Westeros 2004      389788.          389788.
## 8  Westeros 2005      465321           389788.
## 9  Narnia  2002       432156           432156
## 10 Narnia  2003       432156           432156
## 11 Narnia  2004       432156           432156
## 12 Narnia  2005       432156           432156
## 13 Gondor  2002       531426           387968.
## 14 Gondor  2003       321465           387968.
## 15 Gondor  2004       235461           387968.
## 16 Gondor  2005       463521           387968.
## 17 Laputa  2002        14235            36179.
## 18 Laputa  2003        34125            36179.
## 19 Laputa  2004        45123            36179.
## 20 Laputa  2005        51234            36179.
```

If you had a pipeline where you had steps between the group_by() call and the mutate() call that sets the mean_income, you have to be careful. Do *not* put an ungroup() in there. The pipeline works because the mean_income is per country when we call mutate(). If we ungrouped, we would get the mean over *all* countries.

```
mean_income |>
    pivot_longer(
        names_to = "year",
        values_to = "mean_income",
        !country
    )                                              |>
```

```
group_by(
    country
)                                                          |>
# Imagine we are doing something here that removes the group...
ungroup()                                                  |>
# The mutate that follows is not grouped so the mean is global...
mutate(
    mean_per_country = mean(mean_income, na.rm = TRUE),
    mean_income = ifelse(
        is.na(mean_income),
        mean_per_country,
        mean_income
    )
)
```

```
## # A tibble: 20 × 4
##    country   year   mean_income   mean_per_country
##    <chr>     <chr>        <dbl>              <dbl>
## 1  Numenor   2002        123456            249185.
## 2  Numenor   2003        132654            249185.
## 3  Numenor   2004        249185.           249185.
## 4  Numenor   2005        324156            249185.
## 5  Westeros  2002        314256            249185.
## 6  Westeros  2003        249185.           249185.
## 7  Westeros  2004        249185.           249185.
## 8  Westeros  2005        465321            249185.
## 9  Narnia    2002        432156            249185.
## 10 Narnia    2003        249185.           249185.
## 11 Narnia    2004        249185.           249185.
## 12 Narnia    2005        249185.           249185.
## 13 Gondor    2002        531426            249185.
## 14 Gondor    2003        321465            249185.
## 15 Gondor    2004        235461            249185.
## 16 Gondor    2005        463521            249185.
## 17 Laputa    2002         14235            249185.
## 18 Laputa    2003         34125            249185.
```

```
## 19 Laputa    2004        45123            249185.
## 20 Laputa    2005        51234            249185.
```

In the last step, we just use pivot_wider() to transform the long table back to the original format; there is nothing special here that you haven't seen in Chapter 5.

This is how we replace all missing data with the mean for the corresponding countries.

## CHAPTER 9

Working with Strings: stringr

The stringr package gives you functions for string manipulation. The package will be loaded when you load the tidyverse package:

```
library(tidyverse)
```

You can also load the package alone using

```
library(stringr)
```

Counting String Patterns

The str_count() function counts how many tokens a string contain, where tokens, for example, can be characters or words. By default, str_count() will count the number of characters in a string.

```
strings <- c(
    "Give me an ice cream",
    "Get yourself an ice cream",
    "We are all out of ice creams",
    "I scream, you scream, everybody loves ice cream.",
    "one ice cream,
    two ice creams,
    three ice creams",
    "I want an ice cream. Do you want an ice cream?"
)
```

© Thomas Mailund 2022
T. Mailund, *R 4 Data Science Quick Reference*, https://doi.org/10.1007/978-1-4842-8780-4_9

```
str_count(strings)
```

```
## [1]  20  25  28  48  55  46
```

For each of the strings in strings, we get the number of characters. The result is the same as we would get using nchar() (but not length() which would give us the length of the list containing them, six in this example).

You can be explicit in specifying that str_count() should count characters by giving it a boundary() option. This determines the boundary between tokens (characters, words, lines, etc.), that is, the units to count.

```
str_count(strings,boundary("character"))
```

```
## [1]  20  25  28  48  55  46
```

If you want to count words instead, you can use boundary("word"):

```
str_count(strings, boundary("word"))
```

```
## [1]  5  5  7  8  9  11
```

You can use two additional options to boundary(): "line_break" and "sentence." They use heuristics for determining how many line breaks and sentences the text contains dependent on a locale().

```
str_count(strings, boundary("line_break"))
```

```
## [1]  5  5  7  8  11  11
```

```
str_count(strings, boundary("sentence"))
```

```
## [1]  1  1  1  1  3  2
```

Notice that the line breaks are not the newlines in the text. The line breaks are where you would be expected to put newlines in your locale().

Finally, str_count() lets you count how often a substring is found in a string:

```
str_count(strings, "ice cream")
```

```
## [1]  1  1  1  1  3 2
```

```
str_count(strings, "cream") # gets the screams as well
```

```
## [1]  1  1  1  3  3  2
```

The pattern you ask `str_count()` to count is not just a string. It is a regular expression. Some characters take on special meaning in regular expressions.[1] For example, a dot represents any single character, not a full stop.

```
str_count(strings, ".")
```

```
## [1]  20  25  28  48  53  46
```

If you want your pattern to be taken as a literal string and not a regular expression, you can wrap it in `fixed()`:

```
str_count(strings, fixed("."))
```

```
## [1]  0  0  0  1  0  1
```

Since the pattern is a regular expression, we can use it to count punctuation characters:

```
str_count(strings, "[[:punct:]]")
```

```
## [1]  0  0  0  3  2  2
```

or the number of times ice cream(s) is at the end of the string:

```
str_count(strings, "ice creams?$")
```

```
## [1]  1  1  1  0  1  0
```

The s? means zero or one s, and the $ means the end of the string.

Or rather, at the end of the string except that it might be followed by a punctuation mark.

```
str_count(strings, "ice creams?[[:punct:]]?$")
```

```
## [1]  1  1  1  1  1  1
```

_____

[1] Regular expressions are beyond the scope of this book, but if you are frequently working with strings, you might want to familiarize yourself with them.

Splitting Strings

Sometimes, you want to split a string based on some separator—not unlike how we split on commas in comma-separated value files. The `stringr` function for this is `str_split()`.

We can, for example, split on a space:

```
strings <- c(
    "one",
    "two",
    "one two",
    "one two",
    "one. two."
)
str_split(strings, " ")
```

```
## [[1]]
## [1] "one"
##
## [[2]]
## [1] "two"
##
## [[3]]
## [1] "one" "two"
##
## [[4]]
## [1] "one" ""    ""    "two"
##
## [[5]]
## [1] "one." "two."
```

Since we are splitting on a single space, we get empty strings for `"one two"` which contains three spaces. If you want any sequence of spaces to work as the separator, you could instead do this:

```
str_split(strings, "[[:space:]]+")
```

```
## [[1]]
```

```
## [1] "one"
##
## [[2]]
## [1] "two"
##
## [[3]]
## [1] "one" "two"
##
## [[4]]
## [1] "one" "two"
##
## [[5]]
## [1] "one." "two."
```

You can use the boundary() function for splitting as well. For example, you can split a string into its words using boundary("word"):

```
str_split(strings, boundary("word"))
```

```
## [[1]]
## [1] "one"
##
## [[2]]
## [1] "two"
##
## [[3]]
## [1] "one" "two"
##
## [[4]]
## [1] "one" "two"
##
## [[5]]
## [1] "one" "two"
```

When we do this, we get rid of the empty strings from the first attempt, and we also get rid of the full stops in the last string.

Capitalizing Strings

You can use the `str_to_lower()` to transform a string into all lowercase.

```
macdonald <- "Old MACDONALD had a farm."
str_to_lower(macdonald)
```

```
## [1] "old macdonald had a farm."
```

Similarly, you can use `str_to_upper()` to translate it into all uppercase.

```
str_to_upper(macdonald)
```

```
## [1] "OLD MACDONALD HAD A FARM."
```

If you use `str_to_sentence()`, the first character is uppercase and the rest lowercase.

```
str_to_sentence(macdonald)
```

```
## [1] "Old macdonald had a farm."
```

The `str_to_title()` function will capitalize all words in your string.

```
str_to_title(macdonald)
```

```
## [1] "Old Macdonald Had A Farm."
```

Wrapping, Padding, and Trimming

If you want to wrap strings, that is, add newlines, so they fit into a certain width, you can use `str_wrap()`.

```
strings <- c(
    "Give me an ice cream",
    "Get yourself an ice cream",
    "We are all out of ice creams",
    "I scream, you scream, everybody loves ice cream.",
    "one ice cream,
    two ice creams,
```

```
      three ice creams",
      "I want an ice cream. Do you want an ice cream?"
)
str_wrap(strings)
```

```
## [1] "Give me an ice cream"
## [2] "Get yourself an ice cream"
## [3] "We are all out of ice creams"
## [4] "I scream, you scream, everybody loves ice cream."
## [5] "one ice cream, two ice creams, three ice creams"
## [6] "I want an ice cream. Do you want an ice cream?"
```

The default width is 80 characters, but you can change that using the width argument.

```
str_wrap(strings, width = 10)
```

```
## [1] "Give me an\nice cream"
## [2] "Get\nyourself\nan ice\ncream"
## [3] "We are all\nout of ice\ncreams"
## [4] "I scream,\nyou\nscream,\neverybody\nloves ice\ncream."
## [5] "one ice\ncream,\ntwo ice\ncreams,\nthree ice\ncreams"
## [6] "I want an\nice cream.\nDo you\nwant an\nice cream?"
```

You can indent the first line in the strings while wrapping them using the indent argument.

```
str_wrap(strings, width = 10, indent = 2)
```

```
## [1] "  Give\nme an ice\ncream"
## [2] "  Get\nyourself\nan ice\ncream"
## [3] "  We are\nall out of\nice creams"
## [4] "  I\nscream,\nyou\nscream,\neverybody\nloves ice\ncream."
## [5] "  one ice\ncream,\ntwo ice\ncreams,\nthree ice\ncreams"
## [6] "  I want\nan ice\ncream. Do\nyou want\nan ice\ncream?"
```

If you want your string to be left, right, or center justified, you can use `str_pad()`. The default is right-justifying strings.

```
str_pad(strings, width = 50)
```

```
## [1] "                         Give me an ice cream"
## [2] "                      Get yourself an ice cream"
## [3] "                    We are all out of ice creams"
## [4] " I scream, you scream, everybody loves ice cream."
## [5] "one ice cream,\n two ice creams,\n   three ice creams"
## [6] "   I want an ice cream. Do you want an ice cream?"
```

If you want to left-justify instead, you can pass `"right"` to the `side` argument.

```
str_pad(strings, width = 50, side = "right")
```

```
## [1] "Give me an ice cream                         "
## [2] "Get yourself an ice cream                     "
## [3] "We are all out of ice creams                  "
## [4] "I scream, you scream, everybody loves ice cream. "
## [5] "one ice cream,\n   two ice creams,\n   three ice creams"
## [6] "I want an ice cream. Do you want an ice cream?    "
```

You need to use `"right"` to left-justify because the `side` argument determines which side to pad, and for left-justified text, the padding is on the right.

If you want to center your text, you should use `"both"`; you are padding both on the left and on the right.

```
str_pad(strings, width = 50, side = "both")
```

```
## [1] "          Give me an ice cream                "
## [2] "          Get yourself an ice cream            "
## [3] "          We are all out of ice creams         "
## [4] " I scream, you scream, everybody loves ice cream. "
## [5] "one ice cream,\n   two ice creams,\n   three ice creams"
## [6] " I want an ice cream. Do you want an ice cream?   "
```

In these padding examples, we do not keep the lengths of the strings below the padding width. If a string is longer than the padding width, it is unchanged. You can use the `str_trunc()` function to cut the width down to a certain value. For example, we could truncate all the strings to width 25 before we pad them:

```
strings |> str_trunc(25) |> str_pad(width = 25, side = "left")
```

```
## [1] " Give me an ice cream"
## [2] "Get yourself an ice cream"
## [3] "We are all out of ice ..."
## [4] "I scream, you scream, ..."
## [5] " one ice cream,\n  two..."
## [6] "I want an ice cream. D..."
```

The `str_trim()` function removes whitespace to the left and right of a string:

```
str_trim(c(
    " one small coke",
    "two large cokes ",
    " three medium cokes "
))
```

```
## [1] "one small coke"   "two large cokes"
## [3] "three medium cokes"
```

It keeps whitespace inside the string.

Since `str_trim()` does not touch whitespace that is not flanking on the left or right, we cannot use it to remove extra spaces inside our string. For example, if we have two spaces between two words, as follows, `str_trim()` leaves them alone:

```
str_trim(c(
    " one small coke",
    "two  large cokes ",
    " three medium cokes "
))
```

```
## [1] "one small coke"  "two  large cokes"
## [3] "three medium cokes"
```

If we want the two spaces to be shortened into a single space, we can use str_
squish() instead.

```
str_squish(c(
    " one small coke",
    "two large cokes ",
    " three medium cokes "
))
```

```
## [1] "one small coke"  "two large cokes"
## [3] "three medium cokes"
```

Detecting Substrings

To check if a substring is found in another string, you can use str_detect().

```
str_detect(strings,"me")
```

```
## [1]    TRUE FALSE FALSE FALSE FALSE FALSE
```

```
str_detect(strings,"I")
```

```
## [1] FALSE FALSE FALSE  TRUE FALSE  TRUE
```

```
str_detect(strings,"cream")
```

```
## [1] TRUE TRUE TRUE TRUE TRUE TRUE
```

The pattern is a regular expression, so to test for ice cream followed by a full stop, you
cannot search for "ice cream.".

```
str_detect(strings, "ice cream.")
```

```
## [1] FALSE FALSE  TRUE   TRUE   TRUE   TRUE
```

You can, again, use a fixed() string.

```
str_detect(strings, fixed("ice cream."))
```

```
## [1] FALSE FALSE FALSE   TRUE FALSE   TRUE
```

Alternatively, you can escape the dot.

```
str_detect(strings,"ice cream\\.")
```

```
## [1] FALSE FALSE FALSE  TRUE FALSE  TRUE
```

You can test if a substring is *not* found in a string by setting the negate argument to TRUE.

```
 str_detect(strings,fixed("ice cream."),negate =TRUE)
```

```
## [1]  TRUE  TRUE  TRUE FALSE  TRUE FALSE
```

Two special case functions test for a string at the start or end of a string:

```
str_starts(strings,"I")
```

```
## [1] FALSE FALSE FALSE  TRUE FALSE  TRUE
```

```
str_ends(strings,fixed("."))
```

```
## [1] FALSE FALSE FALSE  TRUE FALSE FALSE
```

If you want to know where a substring is found, you can use str_locate(). It will give you the start and end index where it found a match.

```
str_locate(strings, "ice cream")
```

```
##      start end
## [1,]    12  20
## [2,]    17  25
## [3,]    19  27
## [4,]    39  47
## [5,]     5  13
## [6,]    11  19
```

Here, you get a start and end index for each string, but string number six has more than one occurrence of the pattern.

```
strings[6]
```

```
## [1] "I want an ice cream. Do you want an ice cream?"
```

You only get the indices for the first occurrence when you use str_locate().

```
str_locate(strings[6], "ice cream")
```

```
##      start end
## [1,]    11  19
```

The function str_locate_all() gives you all occurrences.

```
str_locate_all(strings[6],"ice cream")
```

```
## [[1]]
##      start end
## [1,]    11  19
## [2,]    37  45
```

If you want the start and end points of the strings *between* the occurrences, you can use invert_match().

```
ice_cream_locations <- str_locate_all(strings[6], "ice cream")
ice_cream_locations
## [[1]]
##      start end
## [1,]    11  19
## [2,]    37  45
```

```
invert_match(ice_cream_locations[[1]])
```

```
##      start end
## [1,]     0  10
## [2,]    20  36
## [3,]    46  -1
```

Extracting Substrings

To extract a substring matching a pattern, you can use str_extract(). It gives you the first substring that matches a regular expression.

```
str_extract(strings, "(s|ice )cream\\w*")
```

```
## [1] "ice cream" "ice cream" "ice creams"
```

```
## [4] "scream"     "ice cream" "ice cream"
```

It only gives you the first match, but if you want all substrings that match, you can use str_extract_all().

```
strings[4]
```

```
## [1] "I scream, you scream, everybody loves ice cream."
```

```
str_extract(strings[4], "(s|ice )cream\\w*")
```

```
## [1] "scream"
```

```
str_extract_all(strings[4], "(s|ice )cream\\w*")
```

```
## [[1]]
## [1] "scream"     "scream"     "ice cream"
```

Transforming Strings

We can replace a substring that matches a pattern with some other string.

```
lego_str <- str_replace(strings, "ice cream[s]?", "LEGO")
lego_str
```

```
## [1] "Give me an LEGO"
## [2] "Get yourself an LEGO"
## [3] "We are all out of LEGO"
## [4] "I scream, you scream, everybody loves LEGO."
## [5] "one LEGO,\n  two ice creams,\n  three ice creams"
## [6] "I want an LEGO. Do you want an ice cream?"
```

```
lego_str <- str_replace(lego_str, "an LEGO", "a LEGO")
lego_str
```

```
## [1] "Give me a LEGO"
## [2] "Get yourself a LEGO"
## [3] "We are all out of LEGO"
## [4] "I scream, you scream, everybody loves LEGO."
## [5] "one LEGO,\n  two ice creams,\n  three ice creams"
## [6] "I want a LEGO. Do you want an ice cream?"
```

These two replacement operators can be written as a pipeline to make the code more Tidyverse-y:

```
strings %>%
   str_replace("ice cream[s]?", "LEGO") %>%
   str_replace("an LEGO", "a LEGO")
```

```
## [1] "Give me a LEGO"
## [2] "Get yourself a LEGO"
## [3] "We are all out of LEGO"
## [4] "I scream, you scream, everybody loves LEGO."
## [5] "one LEGO,\n  two ice creams,\n  three ice creams"
## [6] "I want a LEGO. Do you want an ice cream?"
```

Like most of the previous functions, the function only affects the first match. To replace all occurrences, you need str_replace_all().

```
strings %>%
   str_replace_all("ice cream[s]?", "LEGO") %>%
   str_replace_all("an LEGO", "a LEGO")
```

```
## [1] "Give me a LEGO"
## [2] "Get yourself a LEGO"
## [3] "We are all out of LEGO"
## [4] "I scream, you scream, everybody loves LEGO."
## [5] "one LEGO,\n  two LEGO,\n  three LEGO"
## [6] "I want a LEGO. Do you want a LEGO?"
```

You can refer back to matching groups in the replacement string, something you will be familiar with for regular expressions.

```
us_dates <- c(
    valentines = "2/14",
    my_birthday = "2/15",
    # no one knows but let's just go with this
    jesus_birthday = "12/24"
)
```

```
# US date format to a more sane format
str_replace(us_dates, "(.*)/(.*)", "\\2/\\1")
```

```
## [1] "14/2" "15/2" "24/12"
```

The str_dup() function duplicates a string, that is, it repeats a string multiple times.

```
str_c(
    "NA",
    str_dup("-NA", times = 7),
    " BATMAN!"
)
```

```
## [1] "NA-NA-NA-NA-NA-NA-NA-NA BATMAN!"
```

Here, we also used str_c() to concatenate strings. This function works differently from c(); the latter will create a vector of multiple strings, while the former will create one string.

```
# -- concatenation ----------------------------------------
c("foo", "bar", "baz")
```

```
## [1] "foo" "bar" "baz"
```

```
str_c("foo", "bar", "baz")
```

```
## [1] "foobarbaz"
```

A more direct way to extract and modify a substring is using str_sub(). It lets you extract a substring specified by a start and an end index, and if you assign to it, you replace the substring. The str_sub() function is less powerful than the other functions as it doesn't work on regular expressions, but because of this, it is also easier to understand.

If you do not know where a substring is found, you must first find it. You can use str_locate() for this.

```
my_string <- "this is my string"
my_location <- str_locate(my_string, "my")
my_location
```

```
##      start end
```

```
## [1,]    9   10
```

```
s <- my_location[,"start"]
e <- my_location[,"end"]
str_sub(my_string, s, e)
```

```
## [1] "my"
```

```
my_string_location <- str_locate(my_string, "string")
s <- my_string_location[,"start"]
e <- my_string_location[,"end"]
str_sub(my_string, s, e)
```

```
## [1] "string"
```

```
your_string <- my_string
s <- my_location[,"start"]
e <- my_location[,"end"]
str_sub(your_string, s, e) <- "your"
your_string
```

```
## [1] "this is your string"
```

```
your_banana <- your_string
your_string_location <- str_locate(your_string, "string")
s <- your_string_location[,"start"]
e <- your_string_location[,"end"]
str_sub(your_banana, s, e) <- "banana"
your_banana
```

```
## [1] "this is your banana"
```

When you assign to a call to str_sub(), it looks like you are modifying a string. This is an illusion. Assignment functions create new data and change the data that a variable refers to. So, if you have more than one reference to a string, be careful. Only one variable will point to the new value; the remaining will point to the old string. This is not specific to str_sub() but for R in general, and it is a potential source of errors.

```
my_string
```

```
## [1] "this is my string"
```

```
your_string
```

```
## [1] "this is your string"
```

```
your_banana
```

```
## [1] "this is your banana"
```

If you often write code to produce standard reports, virtually the same text each time but with a few selected values changed, then you are going to love str_glue(). This does precisely what you want. You give str_glue() a template string, a string with the mutable pieces in curly brackets. The result of calling str_glue() is the template text but with the values in the curly brackets replaced by what R expression they contain.

The most straightforward use is when the template refers to variables.

```
macdonald <- "Old MacDonald"
eieio <- "E-I-E-I-O"
str_glue("{macdonald} had a farm. {eieio}")
```

```
## Old MacDonald had a farm. E-I-E-I-O
```

The variables do not need to be global. They can also be named arguments to str_glue().

```
str_glue(
    "{macdonald} had a farm. {eieio}",
    macdonald = "Thomas",
    eieio = "He certainly did not!"
)
```

```
## Thomas had a farm. He certainly did not!
```

Generally, you can put R expressions in the curly brackets, and the result of evaluating the expressions will be what is inserted into the template string.

```
str_glue(
    "{str_dup(\"NA-\", times = 7)}NA BATMAN!"
)
```

```
## NA-NA-NA-NA-NA-NA-NA-NA BATMAN!

x <- seq(1:10)
str_glue(
    "Holy {mean(x)} BATMAN!"
    )
## Holy 5.5 BATMAN!
```

## CHAPTER 10

Working with Factors: forcats

If you work with categorical data, you can often represent the categories by strings. Strings have some drawbacks, however. For example, a spelling mistake can easily go undiscovered. If you want your categories ordered, then the lexicographical order on strings will not always be the order you need. As an alternative to encoding categories as strings we have factors. In this chapter, we look at the functionality that the `forcats` package provides for creating and manipulating factors.

The package is loaded when you import `tidyverse`:

```
library(tidyverse)
```

but you can also load it explicitly:

```
library(forcats)
```

Creating Factors

The built-in `factor()` function is still an excellent choice to build a factor from scratch, especially if you want to specify the levels of the factor when you create it.

If you create a factor without explicitly specifying the levels, you get a level per unique element in the input.

```
factor(c("A","C","B"))
```

```
## [1] A C B
## Levels: A B C
```

Otherwise, you get the levels you pass to the `levels` argument.

© Thomas Mailund 2022
T. Mailund, *R 4 Data Science Quick Reference*, https://doi.org/10.1007/978-1-4842-8780-4_10

```
factor(c("A","C","B"),levels =c("A","B","C","D"))
```

```
## [1] A C B
## Levels: A B C D
```

```
factor(c("A","C","B"),levels =c("D","C","B","A"))
```

```
## [1] A C B
## Levels: D C B A
```

If you have a value that is not found in the levels, it will be set to NA.

```
factor(c("A","C","B"),levels =c("A","B"))
```

```
## [1] A     <NA> B
## Levels: A B
```

The factor() function is usually the right choice for creating factors. The function is not generic, however, so if you want a way to translate a data structure you have implemented yourself, one that needs a specific method to do so, then you might want to specialize as_factor() from forcats.

If you specialize as_factor(), you can do anything you want, but on normal vectors, as_factor() works similarly to as.factor() from base R. The two functions differ, though, in how they set levels. If you use as.factor(), the levels will be sorted.

```
f1<-as.factor(c("C","B","B","A","D"))
f1# levels ordered alphabetically
```

```
## [1] C B B A D
## Levels: A B C D
```

With as_factor(), the levels are in the order they appear in the input.

```
f2<-as_factor(c("C","B","B","A","D"))
f2# levels in the order they appear in the input
```

```
## [1] C B B A D
## Levels: C B A D
```

Concatenation

If you have two factors, then you can concatenate them using fct_c(). The levels of the resulting factor are the union of the levels of the input factors. The order of the levels depends on the order of the input factors. The levels in the first factor go first, then the levels in the second. If the factors share levels, then the order is determined by the first vector.

For example, consider again the preceding factors f1 and f2.

f1

```
## [1] C B B A D
## Levels: A B C D
```

f2

```
## [1] C B B A D
## Levels: C B A D
```

They have the same levels, but they have them in different orders. If we concatenate f1 with f2, we get the levels in the order from f1.

fct_c(f1, f2)

```
##    [1] C B B A D C B B A D
## Levels: A B C D
```

If we concatenate f2 with f1, then the levels are ordered as in f2.

fct_c(f2, f1)

```
##    [1] C B B A D C B B A D
## Levels: C B A D
```

Consider now a factor that has levels not found in f1, for example, this:

f3<-as_factor(c("X","Y","A"))

It shares one level, A, with f1, but X and Y are unique to f3. If we concatenate with f1 as the first argument, the new levels are included but after the factors from f1 (and the shared factor A).

fct_c(f1, f3)

```
## [1] C B B A D X Y A
## Levels: A B C D X Y
```

If we concatenate with f3 as the first argument, then f3's levels come first.

```
fct_c(f3, f1)
```

```
## [1] X Y A C B B A D
## Levels: X Y A B C D
```

You can concatenate more than one factor using fct_c(). The rules are the same as for two factors.

```
fct_c(f1, f2, f3)
```

```
##    [1] C B B A D C B B A D X Y A
## Levels: A B C D X Y
```

```
fct_c(f2, f3, f1)
```

```
##    [1] C B B A D X Y A C B B A D
## Levels: C B A D X Y
```

```
fct_c(f3, f1, f2)
```

```
##    [1] X Y A C B B A D C B B A D
## Levels: X Y A B C D
```

If you have several factors that do not have the same levels, but you want them to have (without concatenating them), you can use fct_unify(). It takes a list of factors and gives you a list of factors where they all have as levels the union of the levels of the input.

```
fct_unify(fs =list(f1, f2, f3))
```

```
## [[1]]
## [1] C B B A D
## Levels: A B C D X Y
##
## [[2]]
## [1] C B B A D
## Levels: A B C D X Y
```

```
##
## [[3]]
## [1] X Y A
## Levels: A B C D X Y
```

Projection

It happens that your categorical data is too fine-grained and you want to group categories into larger classes. If so, the function you want is fct_collapse(). It lets you map your existing levels to new levels. Its first argument is a factor, and after that, you provide named arguments. As parameters to the named arguments, you must provide a list of level names. Each name becomes a level, and that level will contain the elements in the list you give as the argument.

```
fct_collapse(
    fct_c(f3, f1),
    a =c("A","X"),
    b =c("B","Y"),
    c =c("C","D")
)
```

```
## [1] a b a c b b a c
## Levels: a b c
```

You do not need to remap all levels. Those you do not map will stay as they were.

```
fct_collapse(
    fct_c(f3, f1),
    a =c("A","X"),
    b =c("B","Y")
)
```

```
## [1] a b a C b b a D
## Levels: a b C D
```

If you only want to rename and not collapse levels, you can use fct_recode().

```
f1
```

```
## [1] C B B A D
## Levels: A B C D
```

```
fct_recode(f1, a = "A", b = "B")
```

```
## [1] C b b a D
## Levels: a b C D
```

```
fct_recode(f1, X = "A", X = "B", Y = "C", Y = "D")
```

```
## [1] Y X X X Y
## Levels: X Y
```

If you want to reduce the number of levels based on how many elements they have, rather than merge levels in an explicitly specified way, then you can use fct_lump(). You can merge levels such that you keep the *n* levels with the most elements, for example, for *n* = 5 and *n* = 2:

```
f <- factor(sample(1:10, 20, replace = TRUE), levels = 1:10)
table(f)
```

```
## f
## 1  2  3  4  5  6  7  8  9  10
## 0  1  2  0  4  2  2  0  4   5
```

```
f |> fct_lump(n = 5, other_level = "X") |> table()
```

```
##
## 3  5  6  7  9  10  X
## 2  4  2  2  4   5  1
```

```
f |> fct_lump(n = 2, other_level = "X") |> table()
```

```
##
## 5  9  10  X
## 4  4   5  7
```

You see more than five and two levels here, even though we called fct_lump() with $n = 5$ and $n = 2$. If there are several levels with the same number of elements, the function doesn't pick random ones to keep. You get all with the highest occurrences.

If you use a negative number, you get the least common categories.

```
f |> fct_lump(n = -2, other_level = "X") |> table()

##
## 1   4   8   X
## 0   0   0   20
```

Instead of picking the number of categories you want, you can require that they contain at least or at most a fraction of the data. For that, you use the prop parameter. To get the levels that contain at least 10% of the data, you use prop = 0.1:

```
f |> fct_lump(prop = 0.1, other_level = "X") |> table()

##
## 5   9   10   X
## 4   4   5   7
```

If you want the categories that contain less than 10% of the data, you use prop as well but with a negative value. To get the categories that contain at most 10%, you use

```
f |> fct_lump(prop = -0.1, other_level = "X") |> table()

##
## 1   2   3   4   6   7   8   X
## 0   1   2   0   2       2 0 13
```

A more straightforward way to map categories is to map a fixed number to a new category. The fct_other() lets you pick a list of categories to keep or drop. If you keep categories, then all the others are put in the same "other" category. If you drop categories, then those you drop are placed in an "other" category. You can name the "other" category using the argument other_level.

```
f |> fct_other(keep = 1:5, other_level = "X")

## [1] 2 X X X 5 X X 3 3 X X 5 X 5 X X X 5 X X
## Levels: 1 2 3 4 5 X
```

```
f |> fct_other(drop = 1:5, other_level = "X")
```

```
## [1] X 9 9 9 X 7 7 X X 6 10 X 10 X 9
## [16] 6 10 X 10 10
## Levels: 6 7 8 9 10 X
```

Sometimes, what you want is even simpler. You have levels that are not found in the actual data. You can get rid of the empty levels using fct_drop().

```
f1
```

```
## [1] C B B A D
## Levels: A B C D
```

```
levels(f1) <- LETTERS[1:10]
f1
```

```
## [1] C B B A D
## Levels: A B C D E F G H I J
```

```
fct_drop(f1)
```

```
## [1] C B B A D
## Levels: A B C D
```

Adding Levels

The opposite of fct_drop() is adding extra levels. Assigning to levels(), as we did earlier, will add levels that are not necessarily in the factor, but it might also rename levels at the same time. The levels are renamed, you are not just adding to them. This happened to f1 earlier. Its original levels, 1, 2, 3, and 4, were replaced with A, B, C, and D when we added the new levels.

With the fct_expand() function, you can add levels that are not in your factor, and any levels that are already in the data will be ignored.

```
f1
```

```
## [1] C B B A D
## Levels: A B C D E F G H I J
```

```
f3
```

```
## [1] X Y A
## Levels: X Y A
```

```
fct_expand(f1, levels(f3))
```

```
## [1] C B B A D
## Levels: A B C D E F G H I J X Y
```

Missing data is not considered missing data by R, and often we can simply ignore it. Sometimes, however, there is information in missing data that we need to consider. Since R doesn't play well with missing data in many statistical tests, for good reasons I might add as there is no consistently right way to do it, it is best to translate it into a category that R *will* work with. You can use fct_explicit_na() to map all missing data into a new category:

```
f2
```

```
## [1] C B B A D
## Levels: C B A D
```

```
fna <- f2
fna[2] <- NA
fna[3] <- NA
fna
```

```
## [1] C    <NA> <NA> A D
## Levels: C B A D
```

```
fna <- fct_explicit_na(fna, na_level = "Missing")
fna
```

```
## [1] C    Missing Missing A    D
## Levels: C B A D Missing
```

Reorder Levels

If you construct a factor with factor(), you can control the order of the levels using the levels argument. If you use as.factor(), your levels will get sorted by name, and if you use as_factor(), they will be sorted in the order that the levels appear in the data. You can map between these orders if you want to, though.

If you have a factor created with factor() without specifying the levels, then the levels will be all elements seen in the input sorted in their natural order.

```
f <- factor(sample(LETTERS[1:5], 10, replace = TRUE))
f
```

```
## [1] C A B D E C B A B B
## Levels: A B C D E
```

The function fct_inorder() will order your levels to match the order that the categories are seen in the data—the order you get from as_factor().

```
fct_inorder(f)
```

```
## [1] C A B D E C B A B B
## Levels: C A B D E
```

The functions for reordering levels can also make the factors ordered.[1]

```
fct_inorder(f, ordered = TRUE)
```

```
## [1] C A B D E C B A B B
## Levels: C < A < B < D < E
```

If you want to sort the levels, you do not need forcats functions. The plain old factor() will do nicely.

```
factor(f, levels = sort(levels(f)))
```

```
## [1] C A B D E C B A B B
## Levels: A B C D E
```

---

[1] Ordered factors are not just factors with the levels sorted. Ordered factors specify that there is an order on the levels. Regardless of how the levels are sorted in a factor that is not ordered, the categories do not have a natural order to them. This is important in some statistical tests.

```
factor(f, levels = sort(levels(f)), ordered = TRUE)
```

```
## [1] C A B D E C B A B B
## Levels: A < B < C < D < E
```

You can order levels by their frequency in the factor using fct_infreq():

```
table(f)
```

```
## f
## A B C D E
## 2 4 2 1 1
```

```
fct_infreq(f)
```

```
## [1] C A B D E C B A B B
## Levels: B A C D E
```

```
f |> fct_infreq() |> table()
```

```
##
## B A C D E
## 4 2 2 1 1
```

```
f |> fct_infreq(ordered = TRUE)
```

```
## [1] C A B D E C B A B B
## Levels: B < A < C < D < E
```

```
f |> fct_infreq() |> fct_rev() |> table()
```

```
##
## E D C A B
## 1 1 2 2 4
```

In the last example, I reversed the levels, so the order is smallest to largest rather than largest to smallest. The function fct_rev() reverses the levels; a frequent use for this is when you plot with an axis given by the factor. There, the order is given by the levels, and you might want it reversed.

## CHAPTER 11

Working with Dates: lubridate

The lubridate package is essential for working with dates and fits well with the Tidyverse. It is not, however, loaded when you import the tidyverse package, so you need to explicitly load it.

```
library(lubridate)
```

Time Points

You can create dates and dates with time-of-day information using variations of the ymd() function. The letters y, m, and d stand for year, month, and day, respectively. With ymd(), you should write your data in a format that puts the year first, the month second, and the day last. The function is very flexible in what it can parse as a date.

```
ymd("1975 Feb 15")

## [1] "1975-02-15"

ymd("19750215")

## [1] "1975-02-15"

ymd("1975/2/15")

## [1] "1975-02-15"

ymd("1975-02-15")

## [1] "1975-02-15"
```

© Thomas Mailund 2022
T. Mailund, *R 4 Data Science Quick Reference*, https://doi.org/10.1007/978-1-4842-8780-4_11

You can permute the y, m, and d letters if the order of year, month, and day is different. Each permutation gives you a parser that will interpret its input in the specified order.

```
dmy("150275")
```

```
## [1] "1975-02-15"
```

```
mdy("February 15th 1975")
```

```
## [1] "1975-02-15"
```

If you want to add a time of the day to your date, you can add an hour, an hour and a minute, or an hour, a minute, and a second by using _h(), _hm(), and _hms() variants of the ymd() functions.

```
dmy_h("15/2/1975 2pm")
```

```
## [1] "1975-02-15 14:00:00 UTC"
```

```
dmy_hm("15/2/1975 14:30")
```

```
## [1] "1975-02-15 14:30:00 UTC"
```

```
dmy_hms("15/2/1975 14:30:10")
```

```
## [1] "1975-02-15 14:30:10 UTC"
```

If you have a time object

```
x <- dmy_hms("15/2/1975 14:30:10")
```

then you can extract its components through dedicated functions:

```
c(day(x), month(x), year(x))
```

```
## [1]   15    2 1975
```

```
c(hour(x), minute(x), second(x))
```

```
## [1] 14 30 10
```

```
c(week(x), # The week in the year
```

```
wday(x), # The day in the week
 yday(x)) # The day in the year
```

```
## [1] 7 7 46
```

These functions have corresponding assignment functions that you can use to modify the components of the time point.

```
minute(x) <- 15
wday(x) <- 42
x
```

```
## [1] "1975-03-22 14:15:10 UTC"
```

Time Zones

When you add a time of day, a time zone is also necessary. After all, we do not know what time a given hour is before we know which time zone we are in. If I tell you that I am going to call you at two o'clock, you can't assume that it is two o'clock in your time zone.[1] Unless you tell the functions otherwise, they will assume UTC is the time zone. You can specify another time zone via the tz argument.

```
dmy_hm(
  "15/2/1975 14:00",
  tz = "Europe/Copenhagen"
)
```

```
## [1] "1975-02-15 14:00:00 CET"
```

You can take a time point in one time zone and move it to another in two different ways. You set the time zone and pretend that the day and time of day were already in this time zone. That is, you can just change the time zone attribute of the object and not touch the time information. You can do this using the function force_tz().

```
force_tz(
```

---

[1] I am in CEST right now. In Denmark, we switch between CEST and CET depending on daylight saving time. I obviously don't know where you are. If we don't know each other's time zones, it's a good chance we won't agree on when two o'clock is.

```
  # This is a date/time in CET
  dmy_hm("15/2/1975 14:00", tz = "Europe/Copenhagen"),
  # It will be moved to GMT
  tz = "Europe/London"
)
```

```
## [1] "1975-02-15 14:00:00 GMT"
```

Even though CET and GMT are different time zones, the force_tz() function keeps the hour at 14:00 and simply updates the time zone.

A much more likely situation is that you want to know at what a time point in one time zone was in another time zone. For example, if I promise to call you at two o'clock in Denmark, and you are in the UK, you can translate the time from my time zone to yours using with_tz().

```
with_tz(
  # This is a date/time in CET
  dmy_hm("15/2/1975 14:00", tz = "Europe/Copenhagen"),
  # This moves it to the same time but in GMT
  tz = "Europe/London"
)
```

```
## [1] "1975-02-15 13:00:00 GMT"
```

Copenhagen is one hour ahead of London, so when we move the time 14:00 from Copenhagen to GMT, we get the hour 13:00.

Time Intervals

If you have two time points, you also have a time interval: the time between the two points. You can create an interval object from two time points using the interval() function.

```
start <- dmy("02 11 1949")
end <- dmy("15 02 1975")
interval(start, end)
```

```
## [1] 1949-11-02 UTC--1975-02-15 UTC
```

The infix operator %--% does the same thing.

```
start %--% end
```

```
## [1] 1949-11-02 UTC--1975-02-15 UTC
```

You can get the start and end points of an interval using int_start() and int_end().

```
int <- interval(start, end)
int
```

```
## [1] 1949-11-02 UTC--1975-02-15 UTC
```

```
int_start(int)
```

```
## [1] "1949-11-02 UTC"
```

```
int_end(int)
```

```
## [1] "1975-02-15 UTC"
```

The start point does not have to be before the end point. You can define an interval that starts after it ends.

```
end %--% start
```

```
## [1] 1975-02-15 UTC--1949-11-02 UTC
```

```
int_start(start %--% end)
```

```
## [1] "1949-11-02 UTC"
```

```
int_start(end %--% start)
```

```
## [1] "1975-02-15 UTC"
```

You can flip an interval using int_flip().

```
int_flip(end %--% start)
```

```
## [1] 1949-11-02 UTC--1975-02-15 UTC
```

The function int_standardize() will flip the interval if the start point comes after the end point but otherwise will leave the interval as it is.

```
int_standardize(start %--% end)
```

189

```
## [1] 1949-11-02 UTC--1975-02-15 UTC
```

```
int_standardize(end %--% start)
```

```
## [1] 1949-11-02 UTC--1975-02-15 UTC
```

The int_length() will give you the length of an interval in seconds.

```
x <- now()
int <- interval(x, x + minutes(1)) # from now and one minute forward
int_length(int) # the length is one minute, so 60 seconds
```

```
## [1] 60
```

```
int <- interval(x, x + minutes(20)) # now and 20 minutes forward
int_length(int) / 60 # Dividing by 60 to get the length in minutes
## [1] 20
```

You can check if a point is in an interval using the %within% operator.

```
ymd("1867 05 02") %within% int
```

```
## [1] FALSE
```

```
ymd("1959 04 23") %within% int
```

```
## [1] FALSE
```

```
x %within% int # start point is inside the interval
```

```
## [1] TRUE
```

```
(x + minutes(20)) %within% int # end point is inside the interval
```

```
## [1] TRUE
```

You can update the start and end points in an interval by assigning to int_start() or int_end():

```
int_start(int) <- dmy("19 Aug 1950")
int
```

```
## [1] 1950-08-19 01:00:00 CET--2022-08-23 11:50:29 CEST
```

```
int_end(int) <- dmy("19 Sep 1950")
int
```

```
## [1] 1950-08-19 01:00:00 CET--1950-09-19 01:00:00 CET
```

You can move the entire interval by a fixed amount. For example, you can move the interval one month forward using

```
int_shift(int, months(1))
```

```
## [1] 1950-09-19 01:00:00 CET--1950-10-19 01:00:00 CET
```

Given two intervals, the int_overlaps() function checks if they overlap.

```
int1 <- interval(dmy("19 Aug 1950"), dmy("19 Sep 1950"))
int2 <- interval(dmy("19 oct 1950"), dmy("25 nov 1951"))
int3 <- interval(dmy("19 oct 1948"), dmy("25 aug 1951"))
int4 <- interval(dmy("19 oct 1981"), dmy("25 aug 2051"))
```

```
# int1 ends before int2
int_overlaps(int1, int2)
```

```
## [1] FALSE
```

```
# int3 starts before int1 but they overlap
int_overlaps(int1, int3)
```

```
## [1] TRUE
```

```
# no overlap, int4 is far in the future compared to int1
int_overlaps(int1, int4)
```

```
## [1] FALSE
```

The function int_aligns() checks if any of the four start/end points are equal. That is, either the start or the end point of the first interval must be equal to at least one of the points in the second interval.

The four intervals we have created earlier do not have shared interval end points.

```
c(
  int_aligns(int1, int2),
  int_aligns(int1, int3),
```

```
  int_aligns(int1, int4)
)
```

```
## [1] FALSE FALSE FALSE
```

We can create intervals that do share end points and test int_aligns():

```
int5 <- interval(int_start(int1), int_end(int1) + years(3))
int6 <- int_shift(int5, -years(3))
int7 <- int_shift(int6, -years(3))
```

```
c(
  int_aligns(int1, int5), # share start
  int_aligns(int1, int6), # share end
  int_aligns(int1, int7) # overlaps but does not share endpoints
)
```

```
## [1] TRUE TRUE FALSE
```

## CHAPTER 12

Working with Models: broom and modelr

There are many models to which you can fit your data, from classical statistical models to modern machine learning methods, and a thorough exploration of R packages that support this is well beyond the scope of this book. The main concerns when choosing and fitting models is not the syntax, and this book is, after all, a syntax reference. We will look at two packages that aim at making a tidy interface to models.

The two packages, broom and modelr, are not loaded with tidyverse, so you must load them individually.

```
library(broom)
library(modelr)
```

broom

When you fit a model, you get an object in return that holds information about the data and the fit. This data is represented in different ways—it depends on the implementation of the function used to fit the data. For a linear model, for example, we get this information:

```
model <- lm(disp ~ hp + wt, data = mtcars)
summary(model)

##
## Call:
## lm(formula = disp ~ hp + wt, data = mtcars)
##
## Residuals:
```

© Thomas Mailund 2022
T. Mailund, *R 4 Data Science Quick Reference*, https://doi.org/10.1007/978-1-4842-8780-4_12

```
##      Min       1Q  Median      3Q      Max
## -82.565 -23.802   2.111  35.731  99.107
##
## Coefficients:
##                Estimate Std. Error t value Pr(>|t|)
## (Intercept) -129.9506    29.1890  -4.452 0.000116
## hp             0.6578     0.1649   3.990 0.000411
## wt            82.1125    11.5518   7.108 8.04e-08
##
## (Intercept) ***
## hp          ***
## wt          ***
## ---
## Signif. codes:
## 0 '***' 0.001 '**' 0.01 '*' 0.05 '.' 0.1 ' ' 1
##
## Residual standard error: 47.35 on 29 degrees of freedom
## Multiple R-squared:  0.8635, Adjusted R-squared:  0.8541
## F-statistic: 91.71 on 2 and 29 DF,  p-value: 2.889e-13
```

The problem with this representation is that it can be difficult to extract relevant data because the data isn't tidy. The broom package fixes this. It defines three generic functions, tidy(), glance(), and augment(). These functions return tibbles. The first gives you the fit, the second a summary of how good the fit is, and the third gives you the original data augmented with fit summaries.

```
tidy(model) # transform to tidy tibble
```

```
## # A tibble: 3 × 5
##   term        estimate std.error statistic p.value
##   <chr>          <dbl>     <dbl>     <dbl>   <dbl>
## 1 (Intercept) -130.        29.2      -4.45  1.16e-4
## 2 hp             0.658      0.165     3.99  4.11e-4
## 3 wt            82.1       11.6       7.11  8.04e-8
```

```
glance(model) # model summaries
```

```
## # A tibble: 1 × 12
##   r.squared adj.r.squared sigma statistic p.value
##       <dbl>         <dbl> <dbl>     <dbl>   <dbl>
## 1     0.863         0.854  47.3      91.7 2.89e-13
## # . . . with 7 more variables: df <dbl>,
## #   logLik <dbl>, AIC <dbl>, BIC <dbl>,
## #   deviance <dbl>, df.residual <int>, nobs <int>

augment(model) # add model info to data

## # A tibble: 32 × 10
##    .rownames        disp   hp   wt .fitted .resid
##    <chr>           <dbl> <dbl> <dbl>  <dbl>  <dbl>
##  1 Mazda RX4         160  110 2.62   158.    2.45
##  2 Mazda RX4 Wag     160  110 2.88   178.  -18.5
##  3 Datsun 710        108   93 2.32   122.  -13.7
##  4 Hornet 4 Drive    258  110 3.22   206.   51.6
##  5 Hornet Sporta. . .  360  175 3.44   268.   92.4
##  6 Valiant           225  105 3.46   223.    1.77
##  7 Duster 360        360  245 3.57   324.   35.6
##  8 Merc 240D         147.  62 3.19   173.  -26.1
##  9 Merc 230          141.  95 3.15   191.  -50.4
## 10 Merc 280          168. 123 3.44   233.  -65.8
## # . . . with 22 more rows, and 4 more variables:
## #    .hat <dbl>, .sigma <dbl>, .cooksd <dbl>,
## #    .std.resid <dbl>
```

The broom package implements specializations for most models. Not all of the three functions are meaningful for all models, so some models only have a subset of the functions. If you want your own model to work with broom, let us call it mymodel, then you have to implement specializations of the functions, tidy.mymodel(), glance.mymodel(), and augment.mymodel(), that are relevant for the model.

modelr

The `modelr` package also provides functionality for fitting and inspecting models and for extracting information about model fits. We start with the latter.

Consider the following example model:

```
# Build a model where variable x can help us predict response y
dat <- tibble(
  x = runif(50),
  y = 15 * x^2 * x + 42 + rnorm(5)
)
# Fit a linear model to the data (even though y is quadratic in x)
model <- lm(y ~ x, data = dat)
tidy(model)
```

```
## # A tibble: 2 × 5
##    term        estimate std.error statistic p.value
##    <chr>          <dbl>     <dbl>     <dbl>   <dbl>
## 1 (Intercep. . .    38.9     0.522      74.5 2.82e-51
## 2 x                 15.1     0.829      18.2 3.67e-23
```

We have fitted a linear model to data that is sampled from a linear model. The details of the data and model do not matter, it is just an example. I have used the `broom` function `tidy()` to inspect the model.

Two functions, `add_predictions()` and `add_residuals()`, extend your data with predictions the model would make from each data row and the residuals for each row.

```
add_predictions(dat, model)
```

```
## # A tibble: 50 × 3
##        x     y  pred
##    <dbl> <dbl> <dbl>
## 1 0.633  45.7  48.5
## 2 0.385  43.4  44.7
## 3 0.566  46.1  47.4
## 4 0.922  53.5  52.8
## 5 0.976  56.0  53.6
## 6 0.933  54.1  53.0
```

```
##   7 0.381   43.4   44.7
##   8 0.256   43.7   42.8
##   9 0.257   42.0   42.8
## 10 0.197   42.2   41.9
## # . . . with 40 more rows
```

```
add_residuals(dat, model)
## # A tibble: 50 × 3
##        x      y  resid
##    <dbl>  <dbl>  <dbl>
##  1 0.633   45.7  -2.76
##  2 0.385   43.4  -1.27
##  3 0.566   46.1  -1.32
##  4 0.922   53.5  0.690
##  5 0.976   56.0   2.37
##  6 0.933   54.1   1.10
##  7 0.381   43.4  -1.24
##  8 0.256   43.7  0.890
##  9 0.257   42.0 -0.797
## 10 0.197   42.2  0.281
## # . . . with 40 more rows
```

Predictions need not be for existing data. You can create a data frame of explanatory variables and predict the response variable from the new data.

```
new_dat <- tibble(x = seq(0, 1, length.out = 5))
add_predictions(new_dat, model)
```

```
## # A tibble: 5 × 2
##        x   pred
##    <dbl>  <dbl>
## 1     0    38.9
## 2  0.25    42.7
## 3   0.5    46.5
## 4  0.75    50.2
## 5     1    54.0
```

I know that the x values are in the range from zero to one, but we cannot always a priori know the range that a variable falls within. If you don't know, you can use the seq_range() function to get equidistant points in the range from the lowest to the highest value in your data.

```
seq(0, 1, length.out = 5)
```

```
## [1] 0.00 0.25 0.50 0.75 1.00
```

```
seq_range(dat$x, n = 5) # over the range of observations
```

```
## [1] 0.02272462 0.26437738 0.50603015 0.74768292
## [5] 0.98933569
```

If you have two models and want to know how they compare with respect to their predictions, you can use gather_predictions() and spread_predictions():

```
# comparing line to a (better) model y ~ x^2 + x + 1
model2 <- lm(y ~ I(x^2) + x, data = dat)
gather_predictions(new_dat, model, model2)
```

```
## # A tibble: 10 × 3
##     model       x  pred
##     <chr>   <dbl> <dbl>
##  1 model   0       38.9
##  2 model   0.25    42.7
##  3 model   0.5     46.5
##  4 model   0.75    50.2
##  5 model   1       54.0
##  6 model2  0       43.2
##  7 model2  0.25    42.3
##  8 model2  0.5     44.2
##  9 model2  0.75    49.0
## 10 model2  1       56.8
```

```
spread_predictions(new_dat, model, model2)
## # A tibble: 5 × 3
##         x model model2
##     <dbl> <dbl>  <dbl>
```

```
## 1  0      38.9  43.2
## 2  0.25   42.7  42.3
## 3  0.5    46.5  44.2
## 4  0.75   50.2  49.0
## 5  1      54.0  56.8
```

They show the same data, just with the tables formatted differently. The names gather and spread resemble the `tidyr` functions `pivot_longer` and `pivot_wider`, but the names are taken from the functions `gather` and `spread`, also from `tidyr`, that are deprecated versions of the pivot functions.

Earlier, we made predictions on new data, but you can, of course, also do it on your original data.

```
gather_predictions(dat, model, model2)
```

```
## # A tibble: 100 × 4
##     model     x     y   pred
##     <chr> <dbl> <dbl>  <dbl>
##  1 model 0.633  45.7   48.5
##  2 model 0.385  43.4   44.7
##  3 model 0.566  46.1   47.4
##  4 model 0.922  53.5   52.8
##  5 model 0.976  56.0   53.6
##  6 model 0.933  54.1   53.0
##  7 model 0.381  43.4   44.7
##  8 model 0.256  43.7   42.8
##  9 model 0.257  42.0   42.8
## 10 model 0.197  42.2   41.9
## # . . . with 90 more rows
```

```
spread_predictions(dat, model, model2)
```

```
## # A tibble: 50 × 4
##        x     y model model2
##    <dbl> <dbl> <dbl>  <dbl>
## 1 0.633  45.7  48.5   46.4
## 2 0.385  43.4  44.7   42.9
## 3 0.566  46.1  47.4   45.2
```

```
##  4 0.922  53.5  52.8    54.1
##  5 0.976  56.0  53.6    55.9
##  6 0.933  54.1  53.0    54.4
##  7 0.381  43.4  44.7    42.9
##  8 0.256  43.7  42.8    42.3
##  9 0.257  42.0  42.8    42.3
## 10 0.197  42.2  41.9    42.2
## # . . . with 40 more rows
```

If you have the original data, you can also get residuals.

```
gather_residuals(dat, model, model2)
```

```
## # A tibble: 100 × 4
##    model     x      y  resid
##    <chr> <dbl>  <dbl>  <dbl>
##  1 model 0.633  45.7  -2.76
##  2 model 0.385  43.4  -1.27
##  3 model 0.566  46.1  -1.32
##  4 model 0.922  53.5   0.690
##  5 model 0.976  56.0   2.37
##  6 model 0.933  54.1   1.10
##  7 model 0.381  43.4  -1.24
##  8 model 0.256  43.7   0.890
##  9 model 0.257  42.0  -0.797
## 10 model 0.197  42.2   0.281
## # . . . with 90 more rows
```

```
spread_residuals(dat, model, model2)
```

```
## # A tibble: 50 × 4
##        x      y  model model2
##    <dbl>  <dbl>  <dbl>  <dbl>
##  1 0.633  45.7  -2.76  -0.713
##  2 0.385  43.4  -1.27   0.500
##  3 0.566  46.1  -1.32   0.938
##  4 0.922  53.5   0.690 -0.552
##  5 0.976  56.0   2.37   0.0826
```

```
##   6 0.933  54.1  1.10  -0.348
##   7 0.381  43.4 -1.24   0.505
##   8 0.256  43.7  0.890  1.39
##   9 0.257  42.0 -0.797 -0.275
## 10 0.197  42.2  0.281 -0.0587
## # . . . with 40 more rows
```

Depending on the type of data science you usually do, you might have to sample to get empirical distributions or to split your data into training and test data to avoid overfitting. With modelr, you have functions for this.

You can build *n* data sets using bootstrapping with the function bootstrap().[1]

```
bootstrap(dat, n = 3)
```

```
## # A tibble: 3 × 2
##   strap              .id
##   <list>             <chr>
## 1 <resample [50 x 2]> 1
## 2 <resample [50 x 2]> 2
## 3 <resample [50 x 2]> 3
```

It samples random data points and creates n new data sets from this. The resulting tibble has two columns; the first, strap, contains the data for each sample and the second an id.

The crossv_mc() function—Monte Carlo cross-validation—creates cross-validation data, that is, it splits your data into training and test data. It creates n random data sets divided into test and training data.

```
crossv_mc(dat, n = 3)
```

```
## # A tibble: 3 × 3
##   train              test               .id
##   <list>             <list>             <chr>
## 1 <resample [39 x 2]> <resample [11 x 2]> 1
## 2 <resample [39 x 2]> <resample [11 x 2]> 2
## 3 <resample [39 x 2]> <resample [11 x 2]> 3
```

---

[1] There is also a bootstrap() function in broom, but it is deprecated there, so use the modelr function.

By default, the test data is 20% of the sampled data; you can change this using the test argument.

The crossv_kfold and crossv_loo gives you k-fold data and leave-one-out data, respectively.

```
crossv_kfold(dat, k = 3)
```

```
## # A tibble: 3 × 3
##    train               test                .id
##    <named list>        <named list>        <chr>
## 1 <resample [33 x 2]> <resample [17 x 2]> 1
## 2 <resample [33 x 2]> <resample [17 x 2]> 2
## 3 <resample [34 x 2]> <resample [16 x 2]> 3
```

```
crossv_loo(dat)
```

```
## # A tibble: 50 × 3
##    train               test                .id
##    <named list>        <named list>        <int>
##  1 <resample [49 x 2]> <resample [1 x 2]>    1
##  2 <resample [49 x 2]> <resample [1 x 2]>    2
##  3 <resample [49 x 2]> <resample [1 x 2]>    3
##  4 <resample [49 x 2]> <resample [1 x 2]>    4
##  5 <resample [49 x 2]> <resample [1 x 2]>    5
##  6 <resample [49 x 2]> <resample [1 x 2]>    6
##  7 <resample [49 x 2]> <resample [1 x 2]>    7
##  8 <resample [49 x 2]> <resample [1 x 2]>    8
##  9 <resample [49 x 2]> <resample [1 x 2]>    9
## 10 <resample [49 x 2]> <resample [1 x 2]>   10
## # . . . with 40 more rows
```

As an example, say you have sampled three bootstrap data sets. The samples are in the strap column, so we can map over it and fit a linear model to each sampled data set.

```
samples <- bootstrap(dat, 3)
fitted_models <- samples |>
  mutate(
    # Map over all the bootstrap samples and fit each of them
```

```
    fits = strap |> map(\(dat) lm(y ~ x, data = dat))
  )
fitted_models
```

```
## # A tibble: 3 × 3
##    strap                .id   fits
##    <list>               <chr> <list>
## 1 <resample [50 x 2]> 1       <lm>
## 2 <resample [50 x 2]> 2       <lm>
## 3 <resample [50 x 2]> 3       <lm>
```

```
fitted_models$fits[[1]]
```

```
##
## Call:
## lm(formula = y ~ x, data = dat)
##
## Coefficients:
## (Intercept)            x
##       38.46        15.54
```

Then we can map over the three models and inspect them using broom's glance() function:

```
fitted_models$fits |>
  map(glance) |>
  bind_rows()
```

```
## # A tibble: 3 × 12
##    r.squared adj.r.squared sigma statistic  p.value
##        <dbl>         <dbl> <dbl>     <dbl>    <dbl>
## 1      0.871         0.869  1.85      325. 5.07e-23
## 2      0.813         0.809  1.89      209. 4.12e-19
## 3      0.904         0.902  1.58      451. 4.63e-26
## # . . . with 7 more variables: df <dbl>,
## #   logLik <dbl>, AIC <dbl>, BIC <dbl>,
## #   deviance <dbl>, df.residual <int>, nobs <int>
```

If we were interested in the empirical distribution of x, then we could extract the distribution over the bootstrapped data and go from there.

```
get_x <- function(m) {
  tidy(m) |> filter(term == "x") |>
          select(estimate) %>% as.double()
}
fitted_models$fits |> map_dbl(get_x)
```

```
## [1] 15.54085 13.25745 15.50705
```

If you want to compare models, rather than samples of your data, then modelr has support for that as well. You can make a list of the formulae you want to fit. The formulae() function lets you create such a list.

```
models <- formulae(~y, linear = ~x, quadratic = ~I(x^2) + x)
```

The first argument is the response variable (the left-hand side of the formulae), and the remaining arguments are (named) parameters that describe the explanatory variables, the right-hand part of a model formula.

If you call fit_with() with your data, the fitting function to use (here lm()), and the formulae you wish to fit, then you get what you want—a fit for each formula.

```
fits <- fit_with(dat, lm, models)
fits |> map(glance) |> bind_rows()
```

```
## # A tibble: 2 × 12
##    r.squared adj.r.squared sigma statistic   p.value
##        <dbl>         <dbl> <dbl>     <dbl>     <dbl>
## 1      0.873         0.871 1.83      330.   3.67e-23
## 2      0.983         0.983 0.674    1379.   1.86e-42
## # . . . with 7 more variables: df <dbl>,
## #   logLik <dbl>,  AIC <dbl>,  BIC <dbl>,
## #   deviance <dbl>, df.residual <int>, nobs <int>
```

You will find many model quality measures in modelr, for example, root mean square error:

```
fits |> map_dbl(rmse, data = dat)
```

```
##     linear quadratic
## 1.7973686 0.6533399
```

mean absolute error

```
fits |> map_dbl(mae, data = dat)
```

```
##     linear quadratic
## 1.5265158 0.5116268
```

and many more.

Since overfitting is always a problem, you might want to use a quality measure that at least attempts to take model complexity into account. You have some in the glance() function from broom.

```
fits |> map_dbl(~ glance(.x)$AIC)
```

```
##     linear quadratic
## 206.5262   107.3281
```

If at all possible, however, you want to use test data to measure how well a model generalizes. For this, you first need to fit your models to the training data and then make predictions on the test data. In the following example, I have fitted lm(y ~ x) on leave-one-out data and then applied it on the test data. I then measure the quality of the generalization using RMSE.

```
samples <- dat |> crossv_loo()
training_fits <- samples$train |> map(~lm(y ~ x, data = .))
training_fits |> map2_dbl(samples$test, rmse) |> head(10)
```

```
##         1         2         3         4         5
## 2.8223916 1.3057640 1.3454227 0.7254241 2.5129973
##         6         7         8         9        10
## 1.1552629 1.2763827 0.9248123 0.8273398 0.2941659
```

## CHAPTER 13

Plotting: `ggplot2`

The `ggplot2` package contains a vast number of functions for creating a wide variety of plots. It would take an entire book to cover it all—there are already several that cover it—so I cannot attempt this here. In this chapter, I will only try to give you a flavor of how the package works.

The `ggplot2` package is loaded when you load `tidyverse`, but you can always include it on its own using

```
library(ggplot2)
```

The Basic Plotting Components in `ggplot2`

Unlike in R's base graphics, with `ggplot2` you do not create individual plot components by drawing lines, points, or whatever you need onto a graphics canvas. Instead, you specify how your data should be mapped to abstract graphical aesthetics, for example, x- and y-coordinates, colors, shapes, etc. Then you specify how aesthetics should be represented in the graphics, for example, whether x- and y-coordinates should be plotted as scatter plots or lines. On top of this, you can add graphics information such as which shapes points in a scatter plot should have, which colors the color aesthetics maps, and such. You add attributes as separate steps which makes it easy to change a plot. If you want to add a linear regression to your plot, you can do it with a single command; since `ggplot2` already knows which of your data variables are mapped to the x- and y-coordinates, it simply computes the linear regression and adds it to the plot. If you want to plot your data on a log scale, you tell `ggplot2` that the axis should be log-transformed.

At first, `ggplot2` might seem more complicated than basic R's graphics, but you will soon get used to it.

© Thomas Mailund 2022

T. Mailund, *R 4 Data Science Quick Reference*, https://doi.org/10.1007/978-1-4842-8780-4_13

The main components of `ggplot2` are these:

- Data—Obviously, you have data you want to plot.

- Aesthetics—Aesthetics map data to abstract graphical concepts such as x- and y-coordinates, colors, and fills.

- Geometries—Geometries, geometric objects, determine which kind of plot you are making, for example, whether you will get a histogram, a scatter plot, or a boxplot.

- Statistics—Statistics specify how the data should be summarized before plotting. Your data is not always summarized, that is, the statistics can be the identity mapping. A scatter plot doesn't compute a summary for the x- and y-coordinates, but a regression line or a histogram does.

- Scales—Scales specify how the data you mapped to graphical concepts with the aesthetics should actually be placed on a plot. Your x- and y-coordinate data might be measured in meters, but those meters should be mapped to points on your plot. The scales are responsible for this.

- Coordinates—Coordinates allow you to transform the result of scaling your data to plot components. If, for example, you want your plot to show the y-axis on a log scale, then the coordinate transformation does this.

- Faceting—Faceting splits your plot into subplots based on variables in your data.

You create a `ggplot2` plot using the `ggplot()` function. To that object, you add one or more of the components earlier. Nothing happens until you print the graphical object; printing it will make the plot. A typical pattern is to plot it right away and add the components in the same statement in which you create the graphical object.

```
ggplot(. . . ) + . . . components . . .
```

To add components to a plot, you use *addition*. The `ggplot2` package does *not* use pipelines. You often see data piped into the `ggplot()` call, though, but after that, you must remember to add rather than pipe.

A *layer* creates (part of) the graphics you can see. At a minimum, it must consist of data, aesthetics, a statistics (can be the identity), and a geometric object (that might specify the statistics). Your plot must have at least one layer before your plot shows your data.

Adding Components to a Plot Object

The simplest plot we can create is empty. You can create it by calling ggplot() without any arguments.

```
p <- ggplot()
```

You can see what it consists of by calling the summary() function, but most of the information is not relevant here. I will highlight lines relevant to the components we see in this section as we go along.

```
summary(p)

## data: [x]
## faceting: <ggproto object: Class FacetNull, Facet, gg>
##      compute_layout: function
##      draw_back: function
##      draw_front: function
##      draw_labels: function
##      draw_panels: function
##      finish_data: function
##      init_scales: function
##      map_data: function
##      params: list
##      setup_data: function
##      setup_params: function
##      shrink: TRUE
##      train_scales: function
## vars: function
## super:   <ggproto object: Class FacetNull, Facet, gg>
```

To create a plot, you print the plot object. You can do this explicitly:

```
print(p)
```

or just type it into your R terminal:

```
p
```

but in this case, the plot is empty—we didn't add anything to it when we created p—so I haven't shown the result.

Since putting an object at the outermost level in an R script will print an object, ggplot objects are not always assigned to a variable and then plotted later. You can just write the ggplot() object.

```
ggplot()
```

Why would anyone create an empty plot object? You cannot print the empty object. Well, you can, but you will get an empty canvas, so there is not much point to that. You can, however, build up a plot by adding components to it. You can start with the empty plot and add all you want to it in separate commands.

Adding Data

You add data to a plot as an argument to ggplot(). If you want to add data to the empty plot, you will use geometries; see the following example. If you add data in the ggplot() function, then all components you add to the plot later will be able to see the data.

With some random test data, we can create a plot object with associated data.

```
dat <- tibble(
  foo = runif(100),
  bar = 20 * foo + 5 + rnorm(100, sd = 10),
  baz = rep(1:2, each = 50)
)
p <- ggplot(data = dat)
```

If you call summary(p), you can see the line:

```
## data: foo, bar, baz [100x3]
```

at the top. It shows you the variables in the data. They are not mapped to any graphical objects yet; that is the purview of aesthetics.

Adding Aesthetics

What we see in a plot are points, lines, colors, etc. To create these plots, ggplot2 needs to know which variables in the data should be interpreted as coordinates, which determines line thickness, which determines colors, and so on. Aesthetics do this.

Consider this plot object:

```
p <- ggplot(data = dat, aes(x = foo, y = bar, color = baz))
```

Here, we have specified that foo determines the x-coordinate, bar the y-coordinate, and baz the color. If you check the summary of the plot, you can see the mapping from data variables to graphical objects below the data line:

```
## data: foo, bar, baz [100x3]
## mapping:  x = ~foo, y = ~bar, colour = ~baz
```

Plotting (printing) this will give you an empty plot where the x- and y-axes match the range of the data's x and y (foo and bar) values. The plot is otherwise empty because it does not have a geometry.

Adding Geometries

Geometries specify the type of the plot. They use the aesthetics' maps from the data to graphical properties and create a plot based on them.

One of the most straightforward plots is a scatter plot. We can add the geom_point() geometry to the plot we created earlier to get a scatter plot.

```
p <- ggplot(data = dat, aes(x = foo, y = bar, color = baz)) +
      geom_point()
```

If you call summary(p), you will see these lines at the bottom of the output:

```
## geom_point: na.rm = FALSE
## stat_identity: na.rm = FALSE
## position_identity
```

They tell you that you have a point geometry and that the statistic that maps from the data to a summary is the identity. We do not make any summary of the data when we plot it as points. You can create the plot by printing the plot object; you can see the result in Figure 13-1.

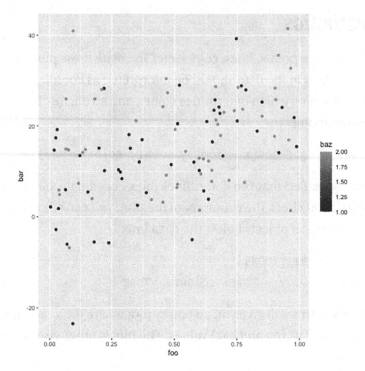

Figure 13-1. *Point geometry plot*

```
print(p)
```

You can add data and aesthetics directly to the geometry. This is especially useful if you want to overlay alternative data onto a plot, but as an example, consider moving the same data and aesthetics to the geom_point() call.

```
p <- ggplot() +
  geom_point(data = dat, aes(x = foo, y = bar, color = baz))
```

If you look at the summary, you will see that the mapping from the aesthetics is now grouped with the geometry:

```
## mapping: x = ~foo, y = ~bar, colour = ~baz
## geom_point: na.rm = FALSE
## stat_identity: na.rm = FALSE
## position_identity
```

but otherwise, there is little change. This can be used to add additional data to a plot. It also shows that it sometimes can make sense to start with an empty plot object and then add different data sets in different geometries.

Since baz is numerical, it is interpreted as a continuous variable. You can get a discrete color mapping by transforming it into a factor.

```
ggplot(data = dat, aes(x = foo, y = bar, color = factor(baz))) +
  geom_point()
```

You can see the result in Figure 13-2.

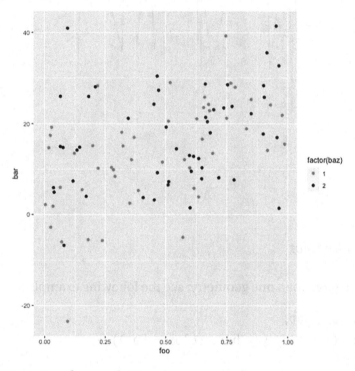

Figure 13-2. *Discrete color aesthetics*

You can change the levels in the factor to reorder the legend. See Chapter 10 for more on manipulating factors.

Another simple geometry is a line plot.

```
ggplot(data = arrange(dat, foo),
       aes(x = foo, y = bar, color = factor(baz))) +
  geom_line()
```

See the result in Figure 13-3. I sorted the data with respect to the x-axis before I plotted it using arrange(). Otherwise, the lines would not go left to right but jump back and forth (try it out if you want to see what I mean).

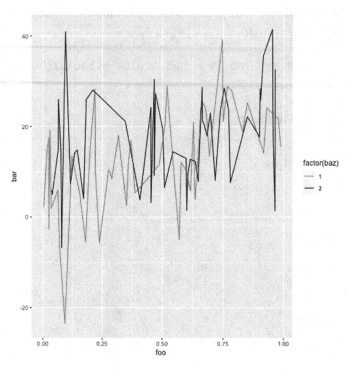

Figure 13-3. *A line plot*

You can have more than one geometry; see the following example.

```
p <- ggplot(data = dat,
            aes(x = foo, y = bar, color = baz)) +
  geom_point() +
  geom_smooth(formula = y ~ x, method = "loess")
```

If you call summary(p), you will see two layers at the bottom of the output.

```
## geom_point: na.rm = FALSE
## stat_identity: na.rm = FALSE
## position_identity
##
```

```
## geom_smooth: na.rm = FALSE, orientation = NA, se = TRUE
## stat_smooth: na.rm = FALSE, orientation = NA, se = TRUE, method = loess
## position_identity
```

Observe that the smooth geometry does not have the identity statistics. It shows a summary of the data, and the mapping from the data to that summary is handled by a stat_smooth statistics.

We can print the plot:

```
print(p)
```

and see the result in Figure 13-4.

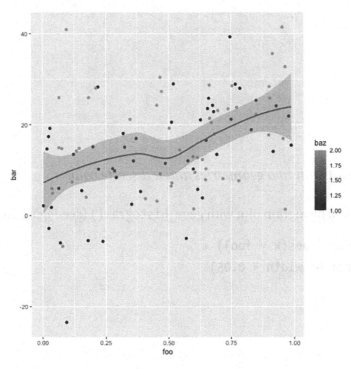

Figure 13-4. *Plot with two geometries*

If we use a discrete color, it also groups the data, so if we make baz a factor again, we will get the data in differently colored point and get two smoothed lines.

```
ggplot(data = dat, aes(x = foo, y = bar, color = factor(baz))) +
  geom_point() +
  geom_smooth(formula = y ~ x, method = "loess")
```

See the result in Figure 13-5.

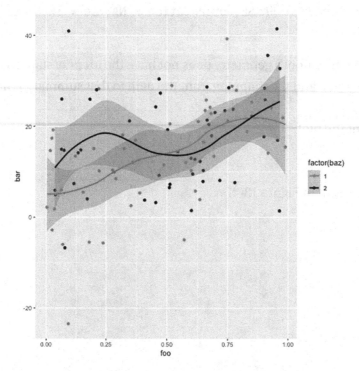

Figure 13-5. *Plot with two geometries and a discrete color*

You can build a histogram plot with geom_histogram() (see Figure 13-6).

```
ggplot(data = dat, aes(x = foo)) +
  geom_histogram(binwidth = 0.05)
```

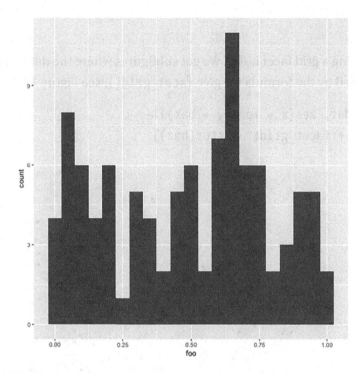

Figure 13-6. *Histogram plot*

Notice that you only need an x-axis for this geometry.

With summary(p), you will see that the statistic is stat_bin.

```
## geom_bar: na.rm = FALSE
## stat_bin: binwidth = NULL, # ...there is more here...
```

If you want a density plot instead, you use geom_density().

```
ggplot(data = dat, aes(x = bar)) +
  geom_density()
```

There, you will see that the statistic is stat_density.

```
## geom_density: na.rm = FALSE
## stat_density: na.rm = FALSE
```

I think you see the pattern now.

Facets

The effect of adding a grid facet is that we get subfigures where the data is split into groups determined by the formula we give `facet_grid()`; see Figure 13-7.

```
ggplot(data = dat, aes(x = foo, y = bar)) +
  geom_point() + facet_grid(~ factor(baz))
```

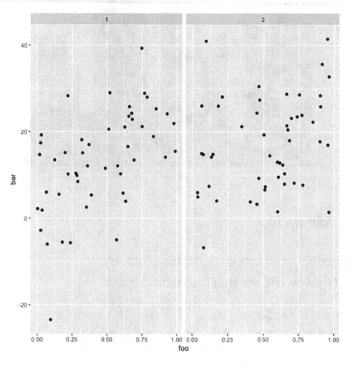

Figure 13-7. *Faceting the plot*

In the previous plots, when you looked at their summary, you would see, under `faceting`, that the class was `FacetNull`.

`## faceting: <ggproto object: Class FacetNull, Facet, gg>`

When we added the grid facet, we now see this:

`## faceting: <ggproto object: Class FacetGrid, Facet, gg>`

You can plot in a two-dimensional grid by having variables on both sides of the formula:

```
dat2 <- tibble(
  foo = rep(1:5, each = 20),
  bar = rep(1:2, each = 50),
  x = foo * bar + rnorm(100),
  y = -foo
)
ggplot(data = dat2, aes(x = x, y = y)) +
  geom_point() + facet_grid(factor(foo) ~ factor(bar))
```

The result is shown in Figure 13-8.

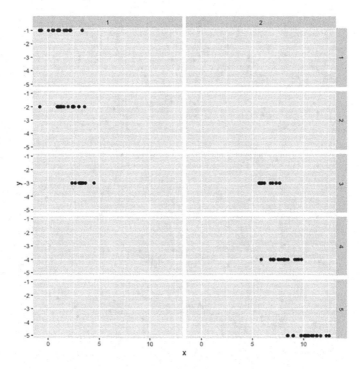

Figure 13-8. *Facet grid for two variables*

You can use more than two variables but (naturally) only two dimensions. If you use more than two variables, then the different categories will be shown as labels on the facet sides. You will see all combinations of factors that appear in the formula. As an example, consider this:

```
dat3 <- tibble(
  foo = factor(rep(1:5, each = 20)),
```

```
  bar = factor(rep(1:2, each = 50)),
  baz = factor(rep(1:5, times = 20)),
  qux = factor(rep(1:2, times = 50)),
  x = rnorm(100),
  y = rnorm(100)
)
ggplot(data = dat3, aes(x = x, y = y)) +
  geom_point() + facet_grid(foo + bar ~ baz + qux)
```

The result is shown in Figure 13-9.

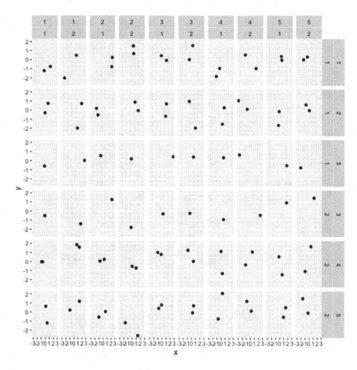

Figure 13-9. *Facet with four variables*

Adding Coordinates

All the preceding plots were plotted in Cartesian coordinates—the default coordinates. You can change the coordinates of a plot. For example, you can flip an axis, such as the x-axis of a plot:

```
ggplot(data = dat, aes(x = foo, y = bar)) +
```

```
geom_point() + coord_flip()
```

Or you can plot in polar coordinates instead of Cartesian coordinates.

```
ggplot(data = dat, aes(x = foo, y = bar)) +
  geom_point() + coord_polar()
```

See Figures 13-10 and 13-11.

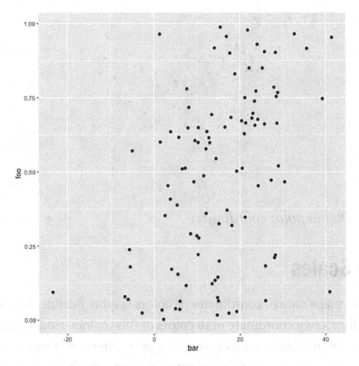

Figure 13-10. *Plot with the x-coordinate flipped*

221

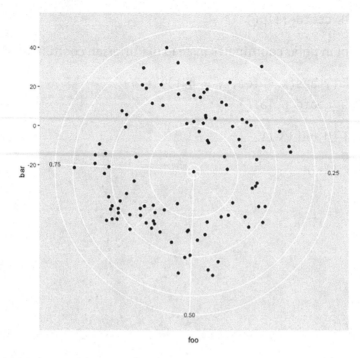

Figure 13-11. *Plot in polar coordinates*

Modifying Scales

The way ggplot2 maps data to coordinates or colors is quite flexible. The aesthetics maps data to either x or y coordinate or to colors or fills (colors used to fill areas), but after that you can modify the corresponding plot properties. The functions for doing this start with scale_, then the property you want to change, and then what you want to do. The operations you modify depend on what you are changing, for example, a coordinate or a color or such.

If you want to put the y axis on a logarithmic scale, for example, you can use scale_y_log10(), and if you want to map a continuous variable to discrete colors, you can use scale_color_binned():

```
ggplot(data = dat, aes(x = bar, y = foo, color = baz)) +
  geom_point() +
  scale_y_log10() +
  scale_color_binned()
```

See Figure 13-12.

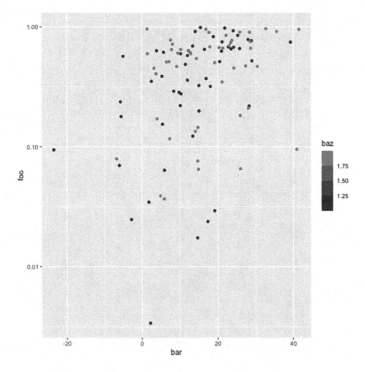

Figure 13-12. *Plot with rescaled y-coordinate and colors*

The list of all these transformation functions is too long to list in this chapter, so I refer you to the package documentation for more information.

With the space available here, I am only able to give you an idea of what you can do with the grammar of graphs implementing in `ggplot2`, but I hope that I have conveyed that with this package you have access to a powerful language for constructing plots. There is much more to it than what I have shown, and I urge you to explore the package in more detail on your own.

CHAPTER 14

Conclusions

R, extended with the Tidyverse, is a powerful language for data science. There is strong support for each step along a data analysis pipeline. After reading this book, you should have a good grasp of how the Tidyverse packages work and how you use them. The book did not cover which data science and machine learning model to use on particular data, but only how they could fit into pipelines. Covering actual data analysis and the methods to use—and packages supporting them—is beyond the scope of this book. Each statistical or machine learning method could fill a book in itself, and there are many such models in R packages. The book is mainly a syntax guide, and what it covered can be used with any model you need to apply to your data, so it is a good foundation for how to adapt your data analysis into the Tidyverse framework.

© Thomas Mailund 2022
T. Mailund, *R 4 Data Science Quick Reference*, https://doi.org/10.1007/978-1-4842-8780-4_14

Index

A
accumulate() function, 104–107

B
Broom packages
    implementation, 193, 194
    tidy(), glance(), and augment()
        functions, 194, 195

C
Column types
    characters, 13
    col_ functions, 20, 22
    function call, 15
    function specification, 21–24
    integer type, 14
    logical type, 17
    ordered argument, 23
    string specification format, 13–21
    time parser, 19
    time point, 18
    type specification, 14, 19
    warning messages, 15

D, E
Data frame manipulation
    arrange() function, 126
    column selection
        column names, 112

        remove columns, 113, 114
        rename() function, 118
        select() function, 112, 116–118
        Sepal.Length, 114, 115
        source code, 111
    desc() function, 126
    dplyr (*see* dplyr functions)
    fictional countries
        dplyr tools, 147
        income, 147
        mean_income, 151
        mutate() function, 149, 150
        pipeline, 148
    filter() function, 119–125
    grouping/summarization
        data-wide summarization, 138
        distinct() function, 137
        group_by() function, 133, 135,
          137, 139
        grouping variables, 134
        group_vars() function, 134
        mutate() function, 136, 137
        summarise() function, 132, 136
        ungroup() function, 136
    joining tables
        anti_join() function, 146
        bind_rows() and bind_columns()
          functions, 139, 140
        distinct() function, 142
        full_join() function, 144
        inner_join() function, 140–143, 145
        left_join()/right_join() function, 144

T, U, V, W, X, Y, Z

Printed in the United States
by Baker & Taylor Publisher Services